6.95

This book is to be returned on or before
the last date stamped below.

17 JAN 1990

KV-046-476

LIVERPOOL POLYTECHNIC LIBRARY

3 1111 00147 8997

Stott, M
The nuclear controversy: a guide to the
T M 621.4838 STO 1980

WITHDRAWN

THE TOWN AND COUNTRY PLANNING ASSOCIATION

The TCPA was one of the leading objectors at the public inquiry into the Nuclear reprocessing plant at Windscale. Founded in 1899 at the Garden City Association, the TCPA is the oldest voluntary body in the world concerning itself with planning and the environment. It is an educational charity supported by a wide ranging membership from all levels of the community and has had considerable influence on planning and environmental policy and practice. It maintains a steady output of informed comment on contemporary planning and environmental issues. Its own evidence to the Windscale inquiry was published in book form on 7 March 1978, entitled *'Planning and Plutonium'*.

Details of Membership and periodical subscriptions are available from:

> The Director
> Town and Country Planning Association
> 17 Carlton House Terrace
> London SW1Y 5AS (Tel: 01—930 8903)

POLITICAL ECOLOGY RESEARCH GROUP

PERG is a public interest science group concerned with environmental issues. Initially an association of university research workers drawn from the social and natural sciences, the group now has a full time staff and is incorporated as a non-profit cooperative. As well as providing consultant expertise for environmental organisations and conducting its own research, the group takes an active part in the policy process by participating in inquiries and hearings (Windscale Inquiry 1977, Brunner Hearings of the European Commission 1978) and lobbying Parliament and Government departments. Recently the Group worked on contract to the State Government of Lower Saxony for the review of the Gorleben reprocessing plant which led to significant changes in licensing requirements. Current research centres on the impact of energy supply technologies on the social and natural environment, as well as upon the political processes of decision making on energy supply policy.

PERG can be contacted at:
> P.O. Box 14
> 34 Cowley Road
> Oxford (Tel: 0865 725354)

The Nuclear Controversy

A guide to the issues of the Windscale Inquiry

Martin Stott and Peter Taylor

THE TOWN AND COUNTRY PLANNING ASSOCIATION
IN ASSOCIATION WITH
THE POLITICAL ECOLOGY RESEARCH GROUP

*Published by the Town and Country Planning Association
17 Carlton House Terrace, London SW1
in association with the Political Ecology Research Group*

© *Martin Stott and Peter Taylor 1980*

First published 1980

All rights reserved. No part of this publication may be reproduced, stored in a retrieval system, or transmitted in any form or by any means, electronic, mechanical, photocopying or otherwise, without the prior permission of the copyright owner.

ISBN 0 902797 04 2

*Design, Typeset and Printed by
Katerprint Co. Ltd. Oxford.*

Contents

Preface	vii
The commercial development of the civil nuclear power programme in Britain	1
The development of the Windscale site	15
The events leading up to the Windscale Inquiry	17
Development and resources of the Inquiry	
Protagonists and their respective positions	20
In favour of development	20
Neutral/seeking specific safeguards	23
Opposed to the development	24
The resources of Inquiry protagonists	33
A brief analysis of relationships between the participant groups	35
Index to issues	41
How to use this guide	44
1. World and UK energy demand	45
2. Weapons proliferation	64
3. Security and civil liberties	72
4. Health and safety	80
5. Conventional planning issues	176
6. Democratic accountability	183
Appendix 1. Inquiry transcript index	191
Appendix 2. A note on British planning inquiry procedure, and its application at the Windscale Inquiry	196
Appendix 3. The terms of reference set by the Secretary of State for the Environment for the Inquiry to be conducted by Mr. Justice Parker, and the questions that arose from them	199
Index	200

Preface

The Windscale Inquiry was the first major examination in public of the nuclear controversy. Although the Inquiry was concerned with outline planning permission for a reprocessing plant designed to handle oxide fuel at a site which already contained a reprocessing plant for 'magnox' fuel, together with various other nuclear installations, the terms of reference of the Inquiry were widened and encompassed many of the issues raised by nuclear power in general. The proceedings were thus expanded and the Inquiry lasted for 100 days. The transcript of the proceedings contains more than 2.5 million words in its 8000 pages, and supporting documents run to several hundred references ranging from single sheets to major technical books.

It is perhaps not surprising therefore that the report of the Inquiry by Justice Parker does not go into this supporting evidence in any detail. The report does not attempt to represent the evidence of both sides, rather the arguments are briefly summarised and reasons given for the position taken by Parker. Some argument and evidence is not presented at all, in particular that relating to alternative energy supplies. In addition to the brevity of the Report and its omissions, it presents a further obstacle to researchers, in that at no point is it referenced to the daily transcript. We felt at the time of the Inquiry that the information then released by British Nuclear Fuels Ltd. (BNFL), together with the documentation used in evidence, and the arguments of the various parties, constituted an important resource for those active in the nuclear debate and for researchers in science policy, sociology of law, environmental politics and so on. When the Parker Report was published its shortcomings were apparent in this and other respects and we were urged by many to proceed with our projected documentation, which at that time was intended to be a working index of the material. We have thus gone further and produced a guide which both acts as an index to the documentation, and as an account of the issues. We have sought to present the material in a neutral tone and in as objective a way possible. As partial observers and participants we recognise that in some areas we may have fallen short of this ideal. For example, we have accorded more space for detail on matters that we felt were inadequately represented in the Parker Report. However, we have sought to include at least a reference to all significant contributions.

The guide sections have been written to aid the specialist researcher who also has access to the Inquiry Transcript and to the full Parker Report (the whereabouts of accessible transcripts can be obtained from the Political Ecology Research Group (PERG), but a full copy is held in the Department of the Environment Library, 2 Marsham Street, London SW1. However, the text has been written in such a way that it will act as a guide to the issues for those who do not wish to pursue matters by reference to the Transcript.

We wish to emphasise that this is very much a working document. We have written in a deliberate style and we recommend that readers approach the document section by section according to the topic of relevance.

The guide has been circulated in draft form to a number of individuals, whose comments have proven invaluable. However, any remaining errors or omissions are the responsibility of the authors, who would appreciate any further comments or suggestions. The authors are prepared to issue supplements to correct or extend the work.

ACKNOWLEDGEMENTS

Support for the writing of this work was widely canvassed. The project would not have got off the ground without the moral support and financial aid of the Town & Country Planning Association and the Elmgrant Trust. Crucial moral support also came from Sheila Oakes, General Secretary of the National Peace Council and further financial aid came from the Joseph Rowntree Charitable Trust with administrative help from the National Council of Social Service. Two token donations were received from the Society for Environmental Improvement and the National Peace Council. Dr Robert Blackith gave a generous personal donation. This moral and financial support over the period of writing helped us through particularly difficult times: readers should be warned! The material in this book is draining of the spirit and should be taken is as small a dose as possible. To those who have supported us, we owe more than the simple writing, and they have our heartfelt thanks. We would also like to thank Sheila Fox and Jan Woodworth for much typing, and our helpful readers of the draft, who have included: Czech Conroy, Robin Grove-White, Mike Flood, David Hall, David Pearce, Joseph Rotblat, Alice Stewart, Colin Sweet, Gordon Thompson and Brian Wynne.

Martin Stott
Peter Taylor
Oxford, October 1979

The Commercial Development of the Civil Nuclear Power Programme in Britain

British involvement in nuclear power was initially for purely military purposes. It was to produce an atomic bomb, in cooperation with the Canadians and Americans, in order to end the Second World War. This was known as the 'Manhattan Project'. In October 1945, the Military Chiefs of Staff recommended to the new Labour administration, under Clement Atlee, that Britain develop an independent nuclear capacity. Accordingly, the British scientists based at Los Alamos, New Mexico, returned to England. They were joined by a smaller group of nuclear scientists, who had been evacuated to Canada, and had participated in the Canadian research programme at the Chalk River reactor near Ottawa, a reactor whose primary concern was research into the generation of power, rather than for military use. These two groups of scientists had experience both in the construction of atomic bombs and of reactors. On 29 January 1946, Atlee appointed Lord Portal as Controller of Atomic Energy, Professor Cockcroft became Director of Research, and Christopher Hinton was made Deputy Controller with special responsibility for the production of fissile materials.

An Atomic Energy Act became law on 6 November 1946, and transferred overall control of the atomic energy project from the Department of Scientific and Industrial Research to the Ministry of Supply, under John Wilmot. Within the Ministry, control of the project was exercised by two bodies, the Atomic Energy Council, which determined overall policy, and the Atomic Energy Technical Committee. Both committees were chaired by Lord Portal.

The cheapest and most effective means of producing an atomic bomb was seen to be by the production of plutonium from a natural uranium reactor, and the design and construction of such a reactor was approved by the government in December 1945. Cockcroft's research group, who were to make the necessary fundamental studies established their headquarters on an airfield site at Harwell near Oxford. Hinton's production group were to organise the supply of metallic uranium and other material that would be needed for research. His group's headquarters were opened in a former munitions factory at Risley in Lancashire on 4 February 1946, with a total staff of 18.

Early in 1946 a special Senate Committee chaired by Senator Robert McMahon studied the whole area of American policy on atomic energy. Its report led to the Atomic Energy Act 1946, commonly known as the McMahon Act, which set up a civilian Atomic Energy Commission (AEC), with ownership and control of all fissionable material in the USA. It also extended the security cover of the atomic weapons programme to include all the activities of the AEC. This effectively prevented any cooperation between Britain and the USA on atomic research, and stopped the import of fissile materials from the USA to Britain. The passing of the Act was a further impetus for the British scientific establishment to press ahead with its own independent research into the development of an atomic bomb. In May 1946, construction of a factory began on the site of a disused poison gas factory at Springfields near Preston to be used as the uranium metal plant filling the gap left by the cessation of imports from the USA. The first bars of uranium to be produced entirely from ore were cast at Springfields on 19 October 1948.

In the meantime Britain's first successful research reactor, the graphite moderate GLEEP, was completed at Harwell, and was taken critical on 15 August 1947. A similar but larger reactor BEPO was also built at Harwell. These plants were used to observe the behaviour of uranium, graphite and other materials in order to confirm theoretical calculations and so provide the basis for further studies. The requirements of military strategy, however, precluded Hinton's proposal that the first reactor programme should consist initially of a simple reactor for producing plutonium alone, followed by a second, more sophisticated plant for both plutonium manufacture and electricity generation; each producing half the country's plutonium requirements. Lord Portal felt that the military programme would not accept the delay in reaching full output which this plan involved. As a consequence two simple reactors were built on the site of a wartime TNT factory at Sellafield in Cumberland. To avoid confusion with the Springfields factory, the site was officially renamed Windscale and site clearance began in September 1947. Construction of the reactors commenced a mere two months later. The first reactor at the site was taken critical in October 1950; the second in June 1951.

Sufficient plutonium had been produced from these reactors for the first phase of the British atomic energy programme to reach a successful conclusion, with the detonation of Britain's first atomic bomb on 3 October 1952 at the Monte Bello Islands site off North-Western Australia. Throughout this time the only indication that the British public had of the programme was a passing reference to it in the House of Commons on 12 May 1948: 'Research and development continues to receive the highest priority in the defence field, and all types of weapons, including atomic weapons, are being developed'.

The development of the 'Cold War', and the return of a Conservative government under Winston Churchill in 1951, led to a shift in emphasis in the nuclear programme. The British government, after consultations with the American government, agreed to abandon its plans for a third Windscale-type pile to increase plutonium output. Instead it was decided to build an isotope separation plant that would convert natural uranium into uranium greatly enriched with the fissile isotope U-235. An old ordnance factory at Capenhurst, Cheshire, had been acquired for the plant, and construction began there in 1949. By 1951 the plant's objectives had been modified and extended to make it a high separation process with an output of almost pure U-235. This, combined with a decision to construct a reactor that would produce both plutonium and electrical power, PIPPA, (pressurised pile for producing power and plutonium) meant that by early 1953 the purely military atomic programme had been modified to one in which the commercial generation of electrical power began to assume some significance.

On 30 January 1953, Duncan Sandys, Minister of Supply, announced Britain's first nuclear power programmes in the White Paper 'A Programme of Nuclear Power'. One was to be the construction of a full-scale nuclear power station, the other a long-term research study into the possibility of developing a fast breeder reactor. In mid-March, the Ministry confirmed that a nuclear power station was to be built immediately adjacent to the Windscale site and that this electricity generating station would be known as Calder Hall. Work immediately began at Risley to modify the PIPPA reactor design for use at Calder Hall, and work on the site started in August 1953. With the move away from the purely military utilisation of nuclear fission, suggestions were made (particularly by Lord Cherwell) that atomic energy should be controlled by an independent body on the lines of the US Atomic Energy Commission, rather than by the Ministry of Supply. In April 1953, Sir Winston Churchill announced the setting up of a committee to evolve a plan for the transfer of responsibility from the Ministry of Supply to an independent body controlling atomic energy. The Committee's main recommendations were published in November 1953 as a White Paper entitled 'The Future Organisation of the United Kingdom Atomic Energy Project'. A bill embodying these recommendations was presented to Parliament in February 1954. The Atomic Energy Act became law on 4 June 1954, and the United Kingdom Atomic Energy Authority (UKAEA) came into existence on 19 July, taking over responsibility for existing establishments on 1 August. The staff employed by the Authority were organised in three main groups: the Weapons Group, under Sir William Penney at the Atomic Weapons Research Establishment at Aldermaston: the Research Group under Sir John Cockcroft at Harwell; and the Industrial Group under Sir Christopher Hinton at Risley, the latter also having responsibility for the Springfields, Capenhurst, Windscale and Calder Hall sites. The

Research Group had responsibility for the Radio Chemical Centre at Amersham.

In March 1954, the Ministry of Works announced that an experimental fast reactor would be built at Dounreay near Thurso in northern Scotland.

Construction of the Dounreay Fast Reactor (DFR), designed to produce 14 MW(e) of power, began in March 1955, and this reactor was taken critical on 14 November 1959. It was taken to its designed full power in July 1963, at which time it began to supply electricity to the north of Scotland Hydro-electric Board.

In June 1955, with the construction work on the two reactors at Calder Hall well advanced, the Ministry of Fuel and Power announced that a further two, to be known collectively as Calder Hall 'B', would be built alongside the existing Calder Hall 'A' station; and shortly after announced that a second station, with four reactors would also be constructed at Chapelcross in Dumfriesshire, Scotland. The first of the Calder Hall reactors was completed in May 1956, with loading of the fuel starting on 17 May and being completed on 22 May. The official opening, by H.M. the Queen, was on 17 October 1956.

A little earlier, in February 1955, the Ministry of Fuel and Power had published a second White Paper entitled, 'A Programme for Nuclear Power'. The programme outlined was to include four stations similar in design to Calder Hall. Construction of the first two was to start in 1957, so that they could be brought into operation in 1960–61, and the third and fourth stations would follow about 18 months later. Each of these four stations would have two reactors with an output of 50–100 MW(e) each (i.e. a similar rating to that of Calder Hall), so that by 1963 between 400 and 800 MW(e) of nuclear generating capacity would be available. A further four stations would be started in 1961–62 for operation in 1965. In all the programme envisaged 12 nuclear power stations with a total capacity of 1400–1800 MW(e) by 1965. The general drift of the White Paper was an emphasis on a nuclear power programme providing for an expansion in electricity consumption without a corresponding increase in coal consumption. The economic cost had not been a factor in the military development of the nuclear power programme, but the generating authorities hoped that ultimately the nuclear stations would be generating electricity at a lower cost than coal-fired power stations.

The contracting for these power stations was to be on a 'turnkey' basis, i.e. a single contractor would be responsible for the whole of the design and construction of a station. Thus, on the advice of the UKAEA and with the approval of the Ministry of Fuel and Power, the stations were to be built by a number of consortia, each based on one of the large turbo-alternator manufacturers, in association with the boilermaker and a civil engineering contractor. Other companies with particular interests, and skills would be able to contribute experience and capital to these consortia. The four initial consortia were:

1. *AEI—John Thompson Nuclear Energy Co.*
   Associated Electrical Industries Ltd.; John Thompson Ltd.; Balfour Beatty & Co. Ltd.; John Laing & Son Ltd.
2. *English Electric—Babcock & Wilcox—Taylor Woodrow Atomic Power Group*
   English Electric Co. Ltd.; Babcock & Wilcox Ltd.; Taylor Woodrow Ltd.
3. *GEC—Simon Carves Atomic Energy Group*
   General Electric Co. Ltd.; Simon Carves Ltd.; Motherwell Bridge & Engineering Ltd.; John Mowlem (Scotland) Ltd.
4. *The Nuclear Power Plant Co.*
   C.A. Parsons & Co. Ltd.; A. Reyrolle & Co. Ltd.; Head Wrightson & Co. Ltd.; Sir Robert MacAlpine and Son Ltd.; Whessoe Ltd.; Strachen & Henshaw Ltd., Alex Findley and Co. Ltd.; Clarke Chapman & Co. Ltd.; Parolle Electric Plant Co. Ltd.

A number of these firms had already acted as contractors, or subcontractors for the reactors at Harwell, Windscale, Calder Hall and Chapelcross.

In the autumn of 1955, the Central Electricity Authority announced the sites for its first two nuclear power stations, Bradwell in Essex, and Berkeley in Gloucestershire. The South of Scotland Electricity Board (SSEB) also announced its intention to build a station at Hunterston in Ayrshire. All four consortia submitted tenders, AEI—John Thompson Nuclear Energy Co. being awarded the Berkeley contract, the Nuclear Power Plant Co., the Bradwell contract, and GEC—Simon Carves Atomic Energy Group the Hunterston contract. The designs for the three new stations were similar to that of Calder Hall, but with the total output raised from 200 MW(e) to over 500 (MW(e). A fifth consortium was formed in January 1957.

5. *Atomic Power Constructions Ltd.*
   International Combustion Holdings Ltd.; Crompton Parkinson Ltd.; Richardsons, Westgarth & Co. Ltd.; Trollope & Colls Ltd.; Holland & Hanneh; Cubbits.

This was the time of the Suez Crisis, which led to a significant rise in the price of fuel oil, and a temporary but serious oil shortage in 1957. This added a sense of urgency to negotiations involving the governments of Belgium, France, Holland, Italy, Luxembourg, and West Germany in establishing a common nuclear energy policy. The Euratom Treaty between these nations was signed on 1 January 1958, pooling their research programmes, establishing a supply agency, and defining common policies on prices and customs' tarrifs affecting nuclear materials. Around the same time, the Central Electricity Authority was reconstituted as the Central Electricity Generating Board under the 1957 Electricity Act. The new CEGB almost immediately indicated that it would be more selective in its policy when placing orders for nuclear power stations than the old CEA had been.

Confidence in the new technology was shaken later in the year with the occurrence of a nuclear accident, Britain's only serious one to date, the 'Windscale fire'. On Monday, 7 October 1957 the No. 1 plutonium pile at Windscale was shut down for a routine fuel element change. During the morning of the following day, the pile was taken critical for a second time to release Wigner energy from the graphite blocks, a first attempt the previous evening having apparently failed to produce sufficient heat to release the energy. However, it soon became apparent that something was wrong as the temperature rose much more rapidly than expected and continued to rise for two more days. On the Thursday morning it was realised that there had been a failure of fuel element cladding, overheated by the second criticality, which had been prompted by an incorrect assumption that the temperature in the core had fallen before the Wigner energy had been released. In fact the thermocouples recording the core temperatures were not in the hottest part of the core, so giving a misleadingly low reading. The fuel element cladding rupture was confirmed by evidence that the air discharging from the pile stack was carrying radioactive particles into the atmosphere. Henry Davey, the General Manager of the site, had to consider the possible consequences of the accident outside the reactor building and outside the site itself, and accordingly warned the Chief Constable of Cumberland to prepare for a possible emergency. Attempts had been made to control the fire by feeding carbon dioxide into the cooling system, but without success. So at 7.00 a.m. on Friday, 11 October, Davey authorised the use of water by firelighters, though there were considerable risks of an explosion in doing so because of the steam pressure that would build up. The water was kept on for 24 hours, and the fire brought under control. However, although the immediate crisis was over, the problem of the escape of radioactive particles into the atmosphere, and their deposition in the neighbourhood presented itself. On the Saturday afternoon the Health Physics Manager advised that the concentration of radioactive iodine (I-131) in milk samples from cows around Seascale was dangerously high. It was decided to ban the consumption of milk from cows grazing in an area of 36 km^2, around Windscale, this area eventually being increased to 500 km^2, and including about 600 farms. All the milk produced by cows within this area was collected, and tipped into the sea. The ban lasted until 23 November, and in all, just over 3,000,000 gallons were destroyed.

An inquiry into the accident, headed by Sir William Penney, found the No. 1 pile completely unserviceable after the accident, and its recommendation that in future Wigner energy in graphite should be released not by nuclear heating, when the fuel temperature would be higher than the graphite, but by the use of electric heaters, along with other recommended modifications, led the AEA to announce that the cost of conversion (about £500,000) would make the plutonium produced unacceptably expensive, and that the No. 2 pile would be shut down permanently too.

As the number of projects increased, the staff employed by the UKAEA increased too, thus: April 1955: 19,859; April 1956: 23,973; April 1957: 27,290; April 1958: 30,341; April 1959: 35,260. By 1956, the AEA recognised that resources of both money and manpower were spread too thinly over too many projects, and set up a committee on the reactor programme. The committee recommended that in future the reactor research programme should concentrate on the Industrial Group's advanced gas-cooled reactor (AGR) study at Risley, with a second group working at a lower priority on a gas-cooled heavy water reactor. Sir John Cockroft also convinced the committee that work should not be abandoned on the high temperature gas-cooled reactor at Harwell. There being no room for an additional reactor at Harwell, the UKAEA took over a site at Winfrith Heath in Dorset in July 1957. The most significant consequence of the committee's report was the abandonment of the LEO pressurised water reactor effectively bringing to an end the British development of a light water reactor for large-scale power stations. This was in contrast to other industrial countries, particularly the USA, which have based their nuclear power programmes on light water systems. It was also against the advice of private industry who saw an important export market in pressurised water reactors.

In addition to the three nuclear power stations (at Berkeley, Bradwell and Hunterston) already ordered under the nuclear power programme, a fourth, at Hinkley Point in Somerset, was also ordered, an announcement being made by the Central Electricity Authority on 28 November 1956. After a public inquiry in May 1957, it was announced in September that the contract for construction of the fourth station had been awarded to the English Electric–Babcock & Wilcox–Taylor Woodrow Atomic Power Group, the consortium that had missed out on the earlier three contracts. After several setbacks this station, designed to produce 500 MW(e) of power, was taken critical in October 1965. The delay caused the original estimated cost of £55 million to rise eventually to £77.5 million.

In June 1960, Atomic Power Constructions were awarded the contract for the fifth Magnox station, at Trawsfynydd in Merionethshire, with a planned total net output of 500 MW(e). This was followed a month later by the signing of a contract with the Nuclear Power Plant Co. for a station at Dungeness on the Kent coast, and early in 1961 another contract, this time with English Electric—Babcock & Wilcox—Taylor Woodrow Atomic Power Group was signed for the construction of a station with a planned output of 580 MW(e) at Sizewell in Suffolk. Thus the seven commercial Magnox stations under construction in the summer of 1961 represented over 3000 MW(e) of generating capacity.

Meanwhile the AEA's prototype AGR at Windscale, the reactor to succeed Magnox, was taken critical on 9 August 1962, over a year later than scheduled. The joint European high temperature 'Dragon' reactor at Winfrith Heath, inaugurated in April 1960, was taken critical in August 1964. It is interesting that there was no legislation in the UK specifically concerned with the siting and operation of nuclear power stations until the Nuclear Installations (Licencing and Insurances) Act was passed in July 1959. This Act established a Nuclear Installations Inspectorate within the Ministry of Power.

As the nuclear power programme would now be completed in fewer than the originally anticipated 12 stations, because station capacity was frequently 500 MW(e) or more, rather than the previously envisaged 100–200 MW(e), it soon became apparent that there would not be enough work to maintain five separate consortia. Therefore in late 1959, the Nuclear Power Plant Co. and the AEI—John Thompson consortium announced a collaborative agreement for future tendering, and in February the following year merged to form the Nuclear Power Group. This was followed in August 1960 by the merger of Atomic Power Constructions and the GEC—Simon Carves consortia to create the United Power Co., so reducing the five consortia to three, one each for Sizewell, Dungeness and Trawsfynydd. Nuclear reactor construction was proving to be less profitable than anticipated by the consortia, as quality controls cut into profit margins, indeed Sir Kenneth Hague in his 1962 Annual Report as Chairman of Babcock & Wilcox, estimated the total losses of the British nuclear industry to be about £25 million. The Magnox stations were costing much more than anticipated (e.g. costs at Hunterston rose from £37.5 million to £60 million in five years), and in July 1962 Sir Christopher Hinton, now Chairman of the CEGB, told a meeting of the Parliamentary and Scientific Committee that the capital cost of a then modern Magnox station with an output of 500 MW(e) was of the order of £120/kW of generating capacity, a figure itself much lower than the real costs of Berkeley or Bradwell, but still much higher than the £37/kW of generating capacity for a conventional power station built near a coalfield. Pocock comments laconically: 'The Generating Boards were committed to the installation of 500 MW(e) of uneconomic nuclear plant'. A Commons Select Committee estimated that the CEGB was spending £20 million per annum more than they would otherwise have done, as a result of using nuclear rather than conventional power stations. The costs had to be met by the British public either as taxpayers or as electricity consumers. Nevertheless, in 1961 the CEGB put out for tenders for the construction of Magnox stations at Oldbury on the Severn River and at Wylfa in Anglesea. Oldbury was awarded to the Nuclear Power Group, and on the principle of 'buggins turn', the No. 1 reactor at Wylfa was awarded to the United Power Co., and the No. 2 reactor to English Electric—Babcock & Wilcox—Taylor Woodrow Atomic Power Group, Wylfa being the last station in the then current programme. However in May 1963, the CEGB informed the United Power Co. that their design for the No. 1 reactor was unsatisfactory and that they would not be awarded the contract after all; the whole station being entrusted to English Electric—Babcock & Wilcox—Taylor Woodrow Atomic Power Group. This rejection of the Wylfa design was a major setback for the United Power Co., which had failed to secure any significant contracts during its independent existence, all its income coming from work secured by its predecessors, APC and GEC—Simon Carves. As there was no announcement of a second nuclear power programme to follow the Magnox stations, the partnership between GEC—Simon Carves and APC was dissolved. UPC ceased to exist. The first two of the CEGB Magnox nuclear power stations Berkeley and Bradwell were completed and formally commissioned on 5 April 1963, and these were quickly followed by the Windscale AGR in June.

In April 1964, a White Paper, 'The Second Nuclear Power Programme' was published, outlining a second stage. It proposed that further nuclear stations of the AGR type should be built to provide a total generating capacity of 5000 MW(e), to be commissioned between 1970—75. It was considered that four or perhaps five (if one were to be built in Scotland) new stations would be needed. The site of the first was at Dungeness (Dungeness 'B') and Atomic Power Constructions Ltd. were awarded the contract for the first commercial AGR. Their design was in fact a considerable extrapolation beyond the limits of experience, a 30 MW(e) plant (Windscale) being scaled up to provide a 600 MW(e) commercial reactor, and financial and technical problems have dogged the project ever since. The Under-Secretary of State, Alex Eadie, announced in a reply to a question in the House of Commons in December 1977, that it was now hoped to commission the first of the Dungeness 'B' reactors in 'mid-1978'. However, at the time of writing it remains unopened. The other four stations were sited in due course at Heysham, Hartlepool, Hunterston 'B' and Hinkley Point 'B'.

In February 1966, Frank Cousins, the Minister of Technology, announced that a 250 MW(e) prototype fast reactor (PFR) would be built beside the experimental fast reactor (DFR) at Dounreay in northern Scotland. This DFR was taken critical for the first time in March 1974, three years later than the original target completion date, due mainly to problems in the reactor containment vessel roof. In spite of the delays the *Financial Times* described the DFR as being 'close to the ideal project' on economic grounds. The DFR achieved full thermal power in March 1977, and in the same month the experimental fast reactor which had operated for 17 years was shut down permanently, as had been planned. The fuel reprocessing plant associated with the PFR is due to be commissioned at the end of 1979.

With divisions of opinion existing in the nuclear industry, brought to a head largely by the problems of who was to blame for the massive over-run in the costs of the nuclear power programme to date, the House of Commons Select Committee on Science and Technology (appointed in January 1967) chose, as the first subject for its initial investigation, the nuclear reactor programme. The Select Committee's recommendations, published in October 1967, on the organisation and financing of the industry, were drastic: the disbanding of the British Nuclear Export Executive; the returning of the UKAEA to a purely research and development role; the creation of a 'nuclear boiler' company and the creation of a parallel 'nuclear fuel' company; and the phasing out of existence of the current consortia on completion of their contracts. After several months' discussion the government accepted the Select Committee's report, except that it preferred a degree of competition, and therefore invited the Industrial Reorganisation Corporation to assist the three existing consortia and the reactor design teams from the AEA to regroup themselves into two design and construction companies for nuclear steam-raising plant. The formation of the first of the new design and construction companies was announced in October 1968. Based on one of the existing consortia, the three commercial partners were: Babcock & Wilcox: 25% of share capital; English Electric: 25%; Taylor Woodrow: 4%; plus AEA: 20% pending transfer to the proposed fuel company; and Industrial Reorganisation Corp.: 26% temporarily held until the final outcome of the English Electric–GEC merger was known.

The new company started trading on 1 December 1968 under the name of Babcock–English Electric Nuclear Ltd., and was renamed British Nuclear Design and Construction Ltd. in February 1969. Financial penalties incurred as a result of the delays on Dungeness 'B' were on a scale that threatened to cripple APC. The CEGB approached APC's parent companies International Combustion and Fairey Engineering and in February 1969 an accommodation was announced whereby British Nuclear Design and Construction assumed the management of the project. It was agreed that APC would remain in existence on paper only for so long as it took to complete the Dungeness 'B' contract. As it turned out, the delays at Dungeness 'B' were so serious that APC did not even stay in existence until the station was completed. Fairey Engineering agreed to take over the work at Dungeness that they would have received as subcontractors for APC, while International Combustion had their contract terminated for the manufacture of the boilers, a new contract with Babcock & Wilcox being negotiated. The whole share capital of APC was acquired by the CEGB and transferred to BNDC, who would finish the work.

With the collapse of APC, only the Nuclear Power Group remained as the basis of the second design and construction organisation. When the subsidiaries of GEC had withdrawn, Reyrolle-Parsons remained as the principal representative of the heavy electrical industry among the parent companies. The shareholdings in the revised group were: Reyrolle-Parsons: 20%; Sir Robert MacAlpine: 15%; Clarke Chapman: 10%; John Thompson: 10%; Head Wrightson: 5%; Strachen & Henshaw: 5%; AEA: 20%; Industrial Reorganisation Corp.: 10%—with the AEA's interest to be transferred to the new fuel company on its establishment. The final stage of the reorganisation, the formation of a single commercial fuel company, was delayed by the 1970 general election, British Nuclear Fuels Ltd. (BNFL), and the Radiochemical Centre Ltd. coming into existence on 1 April 1971 under the terms of the 1971 Atomic Energy Act. Although the new Conservative government hoped for private investment in BNFL (with the state retaining a controlling share), the company came into existence as a wholly state owned concern.

Spring 1973 finally brought the unification of the various elements of the industry into a single British reactor design and construction organisation. The new company was to have Sir Arnold Weinstock's GEC as the major shareholder. The new company and two associated holding companies were registered on 28 June 1973. The main holding company was to be the National Nuclear Corporation, shares in which were: GEC: 50%; UKAEA: 15%; British Nuclear Associates: 35%. British Nuclear Associates, the first of the two associated companies was again a holding company to represent all the commercial organisations other than GEC. The majority of the shares in this group were taken by Babcock & Wilcox, i.e. the corporation was analagous to a consortium of an electrical manufacturer, a boilermaker and a nuclear research organisation. Actual implementation of the corporation's policies would be the responsibility of a third associated company, the Nuclear Power Co.. This was to be a design and construction company with no manufacturing facilities of its own.

Progress on the construction of the five AGR stations in the second nuclear power programme continued to be slow, but the first of the series, Hinkley Point 'B' went critical on 1 February 1976, followed shortly afterwards by Hunterston 'B'. Both were completed three years later than the target completion date of 1973, and both cost much more than the original estimates; £140 million instead of £95 million at Hinkley Point 'B' and £132 million instead of £85 million at Hunterston 'B'. These costs represented increases over the original estimates of 45% and 55% respectively, figures which themselves compare favourably with the other three AGR's in the programme, revised estimates for Heysham showing an increase of 67%, while those for Hartlepool and Dungeness were up by 139% and 214% respectively, a £245 million cost over-run to June 1978 in the case of Dungeness 'B'. Even Sir Arthur Hawkins, Chairman of the CEGB described the second nuclear power programme as 'a catastrophe we must not repeat'.

Nevertheless, In December 1973 the CEGB published its proposals for a third nuclear programme of 18 large power stations, each of about 2400 MW(e) capacity, to be ordered between 1974–83. While by 1972 the British stockpile of plutonium was estimated to be the largest in the world, it was felt by some that a future FBR programme would need to be preceded by this third programme of thermal reactors to provide enough plutonium for the initial fuel charge for an FBR programme. Each power station was to include two pressurised water reactors (PWRs) and the SSEB was known to prefer the steam generating heavy water reactors but was expected to conform with the choice of the larger CEGB, making a possible 26 stations to be built, with as many as 52 PWRs. Considerable doubts about the safety of the PWRs were raised in the press on the basis of American experience, and with the understandable prejudice that existed against AGRs, at the time the only alternative seemed to be the steam generating heavy water reactor (SGHWR), the prototype of which had been constructed at Winfrith Heath. The announcement of the choice of the SGHWR for this third programme was made in July 1974 by the Energy Minister, Eric Varley. However the size of the programme was much smaller than had been expected. Only six reactors, each of 660 MW(e) were to be ordered in the next four years. The reorganisation of the industry to implement this programme had started in 1973 and was complete by early 1975. The Nuclear Power Co. acquired the assets of the former BNDC and TNPG organisations, while the financing of the National Nuclear Corporation was decided in 1976, shareholdings in the company being redistributed as follows: (1) UKAEA: 35%; (2) GEC: 30%; (3) Babcock & Wilcox: 12%; (4) Clarke Chapman: 10% (Clarke Chapman merged with Reyrolle Parsons during 1977 to form Northern Engineering Industries); (5) Taylor Woodrow: 5%; (6) Head Wrightson: 3%; (7) MacAlpine: 2.5%; (8) Whessoe; 2%; (9) Strachan and Henshaw: 0.5%. Through British Nuclear Associates, (3)–(9) together nets 35%.

However, by 1976 two events intervened to reduce the advantages of the SGHWR system. It was discovered that modifications in the design would require the use of imported components such as zirconium tubing, i.e. it would no longer be an all-British project, and Walter Marshall, the Department of Energy Chief Scientist, produced a report suggesting that the safety questions raised about PWRs in 1974 had been satisfactorily answered. Opinion, especially in the CEGB, was swinging away from the SGHWR system, feeling being that AGRs would be the better choice of the nuclear programme was solely to provide new power stations in the UK, or that the PWR should be adapted if the industry were to expand into overseas markets. Eighteen months of politicking ended, temporarily at least, when Tony Benn, the Energy Minister, announced at the end of January 1978 that two (not six as in 1974) new power stations would be built using essentially the previous AGR design.

It was estimated that the stations would cost about £650 million each and that construction would begin in 1980 or 1981, for a target completion date of 1986. In addition, the government announced that it would commission the National Nuclear Corporation to produce a British design of the PWR in case the electricity boards wanted to order one in the 1980s. On the same day (8 February 1978) as the location for these new AGRs was confirmed (Heysham for the CEGB and Torness, Lothian for the SSEB), Roy Berridge, Chairman of the SSEB admitted that the damage caused by a major accident at the Hunterston 'B' AGR in October 1977, when sea-water leaked into the base of the pressure vessel of the No. 2 reactor, would cost the board £15−20 million. This figure has since doubled to £36 million. Problems such as this and the increasing size of its generating plants, with the concomitant greater risk from any failure led the CEGB to raise its margin of spare generating capacity from 20% to 28% in March 1977, an investment decision of considerable significance.

The announcement by Tony Benn that only two power stations were to be ordered precipitated another crisis of confidence in the nuclear industry, reported in the financial press during the Spring of 1978 (for example, *Financial Times*, 27 February and 6 March), with speculation that *another* major reorganisation was likely soon, bringing in several more companies including Rolls Royce and Vickers, with the UKAEA share being replaced by direct BNFL participation. The need for a new structure has been made more urgent by the electrcity supply industries' decision to draft its design-phase contract for the two new nuclear power stations in a way that almost necessitates reorganisation. The chief complaint by the generating boards has been that the complex three-tier structure of the industry prevents them from dealing directly with their contractor, the Nuclear Power Co., which has to refer every significant decision to its parent company and GEC. GEC's supervisory contract expired at the end of 1977 and is currently being extended on a month-by-month basis. A decision on *how* to reorganise the NNC has been delayed by industry opposition to a proposal by Tony Benn that the government take a 51% stake through the National Enterprise Board.

However, Rolls-Royce and Northern Engineering Industries announced in December 1978 that they had launched an Anglo-American venture, to be called RNC (Nuclear) in association with the American company Combustion Engineering, each company having a one-third stake. It was intended that the new consortium would be in a particularly strong position to tender for the proposed British demonstration 1300 MW PWR that the electricity boards hoped to order in the early 1980s, and would be able to take advantage of the export potential of the PWR, using Combustion Engineering's American experience with these reactors. However, the serious accident at the Three Mile Island PWR in Pennsylvania, in March 1979 and the adverse worldwide reaction to the use of the PWR has probably caused this venture to be stillborn.

References

Gowing, M. (1974) *Independence and Deterrence: Britain and Atomic Power 1945—1952*, 2 vols. (London: Macmillan).

Pocock, R.F. (1978) *Nuclear Power: Its Development in the United Kingdom*. (London: Allen and Unwin).

Spencer, Sir Kelvin (1977) Proof of evidence on behalf of the Society for Environmental Improvement to the Windscale Inquiry.

The Development of the Windscale Site

The Windscale site began life in 1939 as a TNT factory, under the name of Sellafield. At the end of the war the government stated its intention not to continue using the factory, and accordingly Courtaulds during 1946/47 were planning a rayon producing plant on the site, but this plan was abandoned when they became aware that it was needed for atomic work. The announcement that Sellafield would be the location of Britain's first plutonium-producing reactors was on 23 July 1947, and clearance began in September 1947, under the Ministry of Works, the site being designed for the construction of two simple air-cooled reactors for the production of plutonium for military use. The programme was planned so tightly that construction of the reactors themselves was started only two months after site clearance began. Construction continued smoothly, despite a significant design change, the addition of filters to the tops of the chimneys, being carried out in November 1948. The construction of the first pile was complete by mid-1950, its experimental loading commenced in July, and it was taken critical in October. The second pile followed shortly afterwards, being taken critical in June 1951, and being commissioned in the following October. At a time of severe rationing, their construction consumed large quantities of building materials including: 400,000 tonnes of high grade concrete; 16,000 tonnes of structural steel; 10,000 tonnes of reinforcement steel; 2,000,000 bricks; 160 km electric cable.

In parallel with the construction of the reactors a chemical separation plant was built to extract plutonium from the fuel rods after irradiation. The first reprocessing complex, separation plant B.204 and supporting facilities, was completed in April 1951, and received its first irradiated fuel in February 1952. A highly active storage plant was commissioned in June 1951, and a plutonium finishing plant in March 1952. The whole complex was transferred to the UKAEA on its creation on 1 August 1954.

The two piles operated normally until October 1957 when a fire in the No. 1 pile led to it being written off, and suggested safety improvements to the No. 2 pile led to that one being closed too because the plutonium produced would have been too expensive. Two reactor systems, Calder Hall and

Chapelcross owned by the UKAEA were at this stage also producing plutonium, as well as electricity which was sold to the national grid. It was clear that the quantity of fuel requiring reprocessing when all the major stations of the generating boards were in operation was going to exceed the capacity of the first separation plant. Construction work on a second separation plant, B.205 began in 1960 and was completed in 1964. It is still in use today, though planning permission was given in 1977 for the plant to be refurbished.

In November 1958, an announcement was made of the intention to build a prototype advanced gas-cooled reactor at Windscale (WAGR). Construction began immediately and the WAGR achieved criticality in August 1962, a year later than originally intended. It was formally opened in June 1963 after achieving full power in the spring. In 1966 work started on the construction of a fabrication facility for mixed uranium-plutonium oxide (MOX) fuel elements for the PFR at Dounreay. This plant was commissioned in 1970. When the second separation plant, B.205, was in full operation, B.204 was shut down and converted for the treatment of oxide fuel like that from the Windscale AGR. The liquors from this pre-treatment 'Head End' plant joined the reprocessing stream in B.205. The 'Head End' plant started operations in 1969, but in September 1973 after only about 80 tons of spent fuel had been treated, the plant was shut down following an accident. An exothermic chemical reaction led to a blow-back which caused a release of beta radiation from the radionuclide ruthenium-106. The plant was immediately closed down and remains out of operation. On 1 April 1971, the Windscale site passed from the hands of the UKAEA to the newly formed British Nuclear Fuels Ltd., though the AEA still owns the Windscale AGR and has some laboratories sharing the site with BNFL, and BNFL was constituted with 100% of its share capital in the ownership of the UKAEA.

References

Pockock, R.F. (1978) *Nuclear Power: Its Development in the United Kingdom* (London: Allen and Unwin).
Scott, A.I. (1977) Proof of evidence on behalf of British Nuclear Fuels Ltd. to the Windscale Inquiry.

The Events Leading up to the Windscale Inquiry

British Nuclear Fuels Ltd. and its French equivalent, Cogema, began negotiations with a consortium of 11 Japanese utilities in October 1974 to reprocess spent nuclear fuel. From these negotiations evolved the plan to build a new thermal oxide reprocessing plant (THORP) at Windscale. Public attention was not aroused until 21 October 1975 when the *Daily Mirror* published a cover story revealing BNFL's negotiations with Japan.

Public concern was aroused, and on 15 January 1976 the Energy Minister, Tony Benn, launched a 'great debate' on nuclear power with a public meeting at Church House, Westminster. Representatives from the nuclear industry, environmental groups, and academics participated.

While BNFL and Cogema's negotiations continued, the government announced on 12 March that it had no objections to these negotiations in principle. Cumbria County Council Planning Committee met on 11 May and announced that any planning permission for the proposed BNFL expansion would not be granted without some prior public consultation. When BNFL formally submitted their planning application to Copeland Borough Council on 25 June, it was referred directly to Cumbria County Council; The application caused some surprise, as in addition to seeking permission to construct the thermal oxide reprocessing plant, BNFL also sought permission to proceed with the less controversial refurbishing of the existing Magnox plant, and the construction of a pilot plant to glassify nuclear waste, along with related ancillary developments.

On 29 September, a public meeting (1) on the BNFL application was held at Whitehaven Civic Centre, just one week after the publication of the Sixth Report of the Royal Commission on Environmental Pollution (Chairman, Sir Brian Flowers) *Nuclear Power and the Environment*. The meeting was chaired by the Chairman of the County Planning Committee, and included representatives of the planning department, BNFL, the nuclear regulatory bodies, local councillors, and environmental groups. After this exercise in public participation Cumbria CC's Planning Committee met on 2 November (2) and considered BNFL's proposals for expansion at Windscale. The Director of

Planning reported that the present development proposals formed a departure from a fundamental provision of the County Development Plan for the following reasons:
- (a) The zoning for industrial use reflected the existing activities on the land, rather than made proposals for development.
- (b) Development on the scale proposed represented such an intensification and expansion of the existing development as to be quite beyond what was in contemplation when the Development Plan was prepared and reviewed.
- (c) The processes involved should not be regarded as 'industrial'. UKAEA development was a special case. (Town and Country Planning (Atomic Energy Establishments Special Development) Order 1954.)

In consequence, the committee unanimously passed the motion 'that this committee being minded to approve, subject to agreement of appropriate conditions, the application by BNFL, the Secretary of State be notified of the application as a departure from a fundamental provision of the County Development Plan, and that the other procedures outlined in the Director of Planning's report be implemented'.' Also minuted at the meeting was the following 'whilst the outcome as detailed in the Director of Planning's report, was reassuring, there were matters for which the County Council were not accepting responsibility, including arrangements for safe transport, security, and the risk of a long term build-up of radio-activity in the sea and in the micro organisms it sustains'. The effect of this meeting was that the BNFL development would be approved if the Secretary of State for the Environment did not 'call in' the application for his consideration, as a departure from the County Development Plan, within twenty one days. Pressure from a number of sources was applied to the Minister, including a petition of 27,000 local people, and by Friends of the Earth, and the Lawyers Ecology Group who pointed out that the County Council Planning Committee were not legally able to approve a development made while refusing to take responsibility for the consequences of it, and that they would take legal action if the Secretary of State did not call the application in. On the 21st and last day of the period allowed, the Secretary of State, Peter Shore, announced that he had asked Cumbria CC for more time to consider the application, and a month later, on 22 December he announced in the House of Commons that there would be a public inquiry into BNFL's application for the THORP. In order to speed up the remainder of the planning application he requested BNFL to withdraw the original application, and resubmit it in three parts; the less contentious two could then be approved, and only the thirt part would be subject to a public inquiry.

Accordingly, BNFL withdrew their original application on 21 January 1977 and re-submitted it in four parts on 1 March. These were:

(1) the thermal oxide reprocessing plant apart from the fuel receipt and storage facility (outline planning permission). This was called in by Peter Shore and was the subject of the inquiry;
(2) the fuel receipt and storage facility (outline planning permission);
(3) a first phase of that facility (outline planning permission);
(4) the extension of an existing oxide fuel storage pond so that oxide fuel from existing overseas contracts could continue to be received (full planning permission).

On 21 April 1977, Cumbria County Council considered the applications, deferred a decision on the second application, granted outline planning permission on the third, and full planning permission on the fourth. Peter Shore, the Minister, called in the application on 25 March for a public inquiry under section 35 of the 1971 Town and Country Planning Act, and on 31 March announced the appointment of Mr. Justice Parker to conduct the Inquiry, with Sir Edward Pochin and Professor Sir Frederick Warner as technical assessors. The Inquiry began on 14 June 1977, in Whitehaven Civic Hall, and finished on 4 November 1977, having sat for 100 days.

While the report was being written considerable pressure for a Parliamentary debate on its findings was being exerted, including the signing of an early-day motion by over 200 MPs of all parties.

Mr. Justice Parker presented his report to the Secretary of State for the Environment on 26 January 1978 and it was published on 6 March. Mr. Shore announced his formal refusal of planning permission at the same time in order to allow Parliamentary debate on the report without having to reopen the Inquiry. The report was debated on 22 March, and again on 15 May when the negative resolution to the Special Development Order, laid by Peter Shore to allow the development to proceed was debated, and defeated (i.e. the development was allowed) by 224 votes to 80. This was the first time that a Special Development Order had been used as a device to allow the debate of a planning application after the public inquiry had closed. In normal circumstances information received by a minister after a public inquiry was ended can be used as the basis for legal action on the grounds that the Secretary of State had not made an impartial decision. Thus, in order to allow Parliamentary debate of Justice Parker's report, the Minister had to refuse the planning application and then, subject to Parliamentary approval, rescind that refusal with a Special Development Order.

Sir John Hill, Chairman of the UKAEA, signed the £600m Japanese contract on 24 May.

Notes
1. The transcript of the public meeting in Cumbria CC document 12.
2. Minutes of the meeting, Cumbria CC document 14. Director of Planning's report to the committee is Cumbria CC document 13 and BNFL document 2.

Development and Resources of the Inquiry Protagonists and Their Respective Positions

1. IN FAVOUR OF THE DEVELOPMENT

British Nuclear Fuels Ltd.

British Nuclear Fuels Ltd. was formed in 1971. The United Kingdom Atomic Energy Authority owns its share capital for the Secretary of State for Energy. Its principal activities are concerned with the provision of nuclear fuel cycle services. Counsel for BNFL, Lord Silsoe, enumerated 16 points summarising their case in his opening statement. The first 13 points relate to the first two points in the Secretary of State's letter setting out Mr. Justice Parker's terms of reference, and the last three to points three to six in his letter. They are: (i) The issues in this case do not depend on the decision whether the UK has an FBR programme, or a predominantly nuclear power system. (ii) The second generation of nuclear power stations, the AGRs are in production, being commissioned or nearing completion. Spent fuel will be arising from these stations and something has to be done with it. We must plan on the basis of at least 3000 tons of spent fuel arising from these stations by 1995. (iii) Reprocessing spent fuel enables the reuse of the uranium and plutonium which are in it, that is to say about 97% of the spent fuel in new fuel elements. A single recycle would add about 30–40% to the power generated from the original uranium ore. (iv) To reprocess that fuel is sensible energy conservation. (v) If the option of developing FBRs is to be retained, reprocessing is essential. (vi) Reprocessing is the only established method of controlling for several decades, the radioactive material in spent fuel, and it readily provides a basis for long-term control and ultimate disposal of the waste. (vii) It makes sound economic sense to reduce the initial financing cost and the unit operating cost by accepting foreign business as well. (viii) The company is confident of obtaining enough business to justify the construction of such a plant. (ix) The Windscale works is an establishment where reprocessing of uranium metal fuel and some oxide has already taken place. The site has been the subject of monitoring and control for a long time and it is the right place for the proposed plant. (x) The technology for such a plant is not novel, and the company can draw on its own experience of reprocessing, and on that of its partners in France and Germany. (xi) The Nuclear Installations Inspectorate takes the view that the proposed plant can be built to high standards of safety, and it sees no reason to oppose the construction of the plant on health and safety grounds. (xii) The National Radiological Protection Board has given the advice that there was no basis for rejecting the overall application on grounds related to the potential public health significance of controlled emissions of radioactive wastes to the environment, from Windscale. (xiii) (a) The denial of reprocessing to ourselves would achieve nothing, because we are a nuclear weapons power, with the technology for the enrichment of uranium, and the recovery of plutonium, anyway. (b) Denial of reprocessing to other countries would place serious pressures upon these countries to reprocess their fuel for themselves. (xiv) Banning reprocessing would not end terrorism or its problems. (xv) No problems are raised which cannot be satisfactorily solved concerning the effects of the proposal on the visual scene, upon infrastructure matters, road, rail, etc. (xvi) In relation to employment the proposals would bring a sizeable number of stable jobs to a special development area. BNFL estimated that the cost of the development would be in the region of £600 million.

Copeland Borough Council

The Council is a statutory local authority, a district within the county of Cumbria, created by the local government reorganisation of 1974. The area contains a population of approximately 70,000 people and includes a large part of the Lake District National Park.

The Council's position was one of support for the application. It was satisfied about the environmental and safety issues involved, and while considering the existing monitoring and controlling activities to be satisfactory associated itself with the County Council's request for further improvements in monitoring systems and techniques. The Council was, however, concerned about the effect of the development on local infrastructures and looked to the Secretary of State in determining the application to bear this issue in mind. It intends to seek to have funds made available from central government sources, and from BNFL to meet the extra needs without detriment to their existing programmes.

Cumbria County Council

The council is a statutory local authority created in the local government reorganisation of 1974. It is also the local planning authority and as such its town and country planning committee considered the application from BNFL which was the subject of the Inquiry. The County Council employed Professor Fremlin as an independent expert to advise on matters of safety, and concluded that it was in favour of the development. They concluded that potential risks posed no real problem, being a matter for general government policy, and for the licensing and authorising authorities, rather than a matter for planning control. It considered that there was no substantial planning objection, that the question of planning permission could be decided irrespective of costs, that if the development were not economic it would not go ahead, and that an environmental impact analysis would not be necessary. The Council did hope that, in the event of the proposal being approved, the government and BNFL would contribute to the costs of improving infrastructure in the area.

Central Electricity Generating Board

The CEGB is a nationalised utility supplying electricity to consumers in England and Wales. Arisings from its present and planned AGR power stations would make up half the input into THORP. The Board fully supported BNFL in the application, and considered that THORP was needed in order to deal with committed AGR arisings. It also considered that reprocessing was the only established means known of dealing with these arisings, rejecting storage at least at this stage.

Electrical Electronic, Telecommunications, and Plumbing Union

The EETPU supported the application by BNFL on the basis of its support of the TUC's energy policy. This policy considers that we must seek as a nation to develop a coordinated energy policy which would encourage the development of all available energy resources, with some of the revenue from North Sea oil being used to finance research into energy saving, and alternative sources of energy, such as wave power, wind power and solar power. The EETPU considered that an essential element of this policy should be the continued development of nuclear energy. As a consequence it is necessary to develop methods of conservation and the improved use of energy, which will require that fuels be used as efficiently as possible, and the reprocessing of nuclear fuels is an important part of this programme. The union also welcomed the fact that the development would provide further opportunities for employment within an area of high unemployment. It considered the plant to be safe, that its operation would result in no curtailment of civil liberties, that there would be no risks if a strike occured, and as a consequence there would be no curtailment in the right to strike.

Ridgeway Consultants

Ridgeway Consultants represented by Dr. Little supported BNFL's application on the basis of considering what was best from the point of view of the national interest, and the conditions necessary for Britain to remain a free and independent country. They argued that a country should be economically viable, and that people should have a sense of right and wrong. They criticised anti-nuclear groups for playing into the hands of subversive organisations and the Soviet Union, and objected to their lobbying, their interconnectedness (see Figure 1) and their purveying of fairy stories which confuse the elected representatives as to the real views of the public.

South of Scotland Electricity Board

The SSEB is a nationalised utility responsible for the supply of electricity in the South of Scotland, with responsibility for a population of four million. The Board generates 30% of its electricity from nuclear power and a further reactor is under construction at Torness, near Edinburgh. The Board supported BNFL's application because it considered that reprocessing must form an integral part of the fuel cycle, and that since there is a universal need to utilise all sources of energy fully, it would be advantageous for the UK to be able to influence matters such as uranium supply and waste storage from a position of knowledge and experience. The Board also considered that the economics of scale resulting from having a larger reprocessing plant, in order to take in foreign fuel, would reduce the cost of electricity to the UK consumer.

2. NEUTRAL/SEEKING SPECIFIC SAFEGUARDS

A few groups participating in the Inquiry were neither in favour nor opposed to the development in principle, but participated in order to seek assurances that, if the development did go ahead, the report would include recommendations for specific safeguards which they felt would be necessary if their specific interests were not to be jeopardised.

Lake District Special Planning Board

The Board is responsible for the administration of planning Acts and associated legislation within the Lake District National Park. It has 27 members, two-thirds appointed by Cumbria County Council, and one-third by the Secretary of State for the Environment. The Board did not express a clearly pro- or anti-position but pointed to three areas where it looked for safeguards in the event of the development proceeding, these being: (a) the monitoring of the rate of water abstraction from the rivers Ehen and Derwent; (b) restrictions on the development of local authority, company and private housing within the park, in order to preserve the character of the villages; and (c) that a transport survey of the anticipated extra traffic associated with the site be carried out, and efforts be made to minimise its effects on the park.

Lancashire and Western Sea Fisheries Joint Committee

The Lancashire and Western Sea Fisheries Joint Committee is a public body set up to protect and develop inshore sea fisheries from Haverigg Point, Cumbria to Cenmaes Head, Pembrokeshire. The Committee consists of councillors from the local authorities concerned, water authorities, inshore fishermen, representatives of the wholesale and retail fish trade, scientists and naturalists appointed by MAFF. Professor Potts, the Committee's witness, said under cross-examination by Lord Silsoe (day 65, p. 45) 'we are not opposing the application, merely making suggestions which we think would protect the interests of the fishing industry in the event of the development proceeding'. The Committee's primary concern was with the unsatisfactory state of the effluent from the existing plant, particularly the discharges of Caesium-134 and 137, which they requested should not exceed 100,000 curies in any one year. Its chief fear was to avoid a situation which it considered a real possibility, that certain parts of the Irish Sea would have to be closed to fishing.

3. OPPOSED TO THE DEVELOPMENT

British Council of Churches

The British Council of Churches is an umbrella organisation for all the Christian Churches in Britain except the Roman Catholics. Its opposition to the Windscale development was based on six elements. (i) The significant degree of disagreement amongst experts, relating to the relevant technical issues. (ii) The problem of the safe long-term disposal of wastes derived from reprocessing. (iii) That the development significantly increased the burden of responsibility, and potential of risk, to our descendants. (iv) The prospect that increased production of plutonium, and the possible consequential stimulus to the development of an energy system based predominantly on plutonium, would lead to unacceptable limitations on civil liberties and may increase the risk of proliferation. (v) The foreclosing of alternative, and less dangerous options by a decision to expand a part of the nuclear energy process which gives prominence to the use of plutonium. (vi) The *relative* insignificance of arguments for the application based on foreign earnings and employment opportunities.

Council for Science and Society

The Council for Science and Society is a registered charity engaged in the study of the social consequences of science and technology, and the publication of its results. The Council based its opposition to the development on arguments that the long-term consequences of a major nuclear power programme in the UK were not sufficiently understood to be fully foreseeable, and that it was not known whether there were technical solutions for the technical problems that have been identified. It considered that decisions involving public judgements were 'political' rather than 'scientific' and that the council did not feel qualified to make those judgements.

Durham County Council

Durham County Council is a statutory local authority, immediately to the east of Cumbria. The County Council's objections to the proposed development were based on three points: that it would lead to increased transport of dangerous substances, some of which might pass through County Durham; the possibility of an accident at Windscale which might occasion the release of high or low level radioactive substances; and the risks associated with the storage of radioactive substances which remain active for thousands of years.

Friends of the Earth

Friends of the Earth Ltd. was founded in Britain in 1970, and advocates a programme for the conservation and rational use of the earth and its resources. Its activities include research, education, publishing and lobbying. It currently has about 6500 members and about 150 local groups. It also has an associated research charity Earth Resources Research. FOE were among the leading activists in calling for a public inquiry. Their objection to the development was based on eight propositions. (i) That the plant is unlikely to work, and that attempts to make it work will result in further heavy expenditure. (ii) It is unnecessary, the alleged savings in expenditure on fuel are unreal, with uranium prices being likely to remain highly competetive with the fast-breeder process. (iii) The information about the proposed Japanese contract was vague and contradictory, and the anticipated profits might prove illusory. (The terms of the Japanese contract were released in the Inquiry (BNFL 179), and showed BNFL's profits to be secure, the Japanese having agreed to charges for reprocessing on a 'cost-plus' basis.) (iv) That BNFL hope to find some way on a commercial scale of glassifying of vitrifying long-term waste, but there is no certainty of success. (v) That plutonium should not be further unnecessarily produced, as the return of separated plutonium to customers who might not be subscribers to the Non-proliferation Treaty, and might use it for weapons manufacture, would create a danger of weapons proliferation even greater in its potential results than plutonium in the hands of terrorists. (vi) If plutonium is to be stored or transported on a greatly increased scale beyond that at present considered necessary for weapons, then a military or quasi-military form of security will have to be introduced with a consequent curtailment of civil liberties. (vii) The reprocessing plant, and the storage of waste introduces unnecessary additional hazards of fire, explosion and accident. (viii) The expenditure is grossly out of proportion with who will, or can usefully be employed in this area, and alternative uses for the domestic financial contribution would create more, and more appropriate, employment, both nationally and locally.

FOE's counsel argued that a delay of ten years could not possibly harm any reasonably foreseeable nuclear programme that the government might wish to undertake. In its closing statement FOE also rejected several of the arguments put forward by BNFL during the Inquiry, namely: that reprocessing is necessary as a form of waste management of spent oxide fuel from existing and projected reactors, that there are considerable economic advantages for the country in the proposed development, and that the separation of further plutonium is necessary in order to keep open the fast breeder option.

Friends of the Earth—West Cumbria

FOE—West Cumbria, a local group of national FOE, was founded in 1974, it currently has about 100 supporters. The group is also a constituent member of the Network for Nuclear Concern, and as such associated itself with the wider national and international arguments including those on health and safety put forward by these two groups, while themselves concentrating on matters of more local concern. Specifically these were: (i) That no significance should be attached to the two local councils' original recommendations of support because as they had not initially considered the THORP proposals could not have distinguished or understood THORPs true implications, that they had no mandate from the electorate to support the application, and that they did not have sufficient information on which to base a sound judgement. (ii) That their witnesses showed that a substantial body of local opposition existed. (iii) That the employment generated by the development would do little to alleviate local unemployment, but on the contrary would draw skilled workers from alternative employment in the area, with detrimental effects to these local firms, and that the employees brought in from outside the area would create extra demand on the local infrastructure, place increased pressure on the very sector of local job vacancies where the problem was greatest at the moment, and that the additional wealth generated would not affect these drawbacks. (iv) That the nuclear developments were perpetuating the historical problems of West Cumbria by causing a drain of labour from other local occupations, stultifying incentive to develop alternative employment in the area and in fact only having a limited designed life. (v) That the whole system of local liaison with BNFL needed to be restructured from top to bottom.

Friends of the Lake District and Cumbrian Naturalists Trust

The Cumbrian Naturalists Trust is a registered charity with 2000 members, which owns, manages and has access agreements to the nature reserves in the county. The groups were concerned about the hazards to wildlife and the natural environment posed by the development, particularly about those arising from radioactive releases. They also requested that due consideration be given to the visual impact of the proposal on the Lake District National Park, and the Cumbrian coastline.

Isle of Man Local Government Board

The Isle of Man Local Government Board, which approximates in its functions to the British Department of the Environment, represented the Isle of Man Government at the Inquiry. The island has a population of about 60,000 and although forming part of the British Isles, is not legally part of the UK, enacting its own legislation for all domestic matters. The UK government remains responsible for defence and foreign affairs. The Isle of Man's case was, according to their counsel Mr. Harper, (day 71, p. 33) not one of blanket anti-nuclear opposition, but rather an attempt to emphasise the unknowns present and future, and to argue that the development was premature. Their general opposition centred around four points: (i) fear for the general safety of the Isle of Man population in case of an accident or sabotage; (ii) the general fear that in practice no clear line of demarcation can be drawn between the peaceful uses of nuclear power, and the potential spread of weapons material which would destroy human life once and for all; (iii) the fear that civil liberties and personal freedom would be at risk in a society dependent on the strict security which would need to accompany further development of nuclear power, and (iv) specific concern over the accumulation of radioactive wastes discharged into the Irish Sea. While associating themselves with the cases of other objectors in other fields, the Board concentrated its own case in two areas, those of planning and the unsuitability of the site for the proposed development, and of the long term effects of radioactivity from low level discharges into the Irish Sea.

Justice

Justice, the British section of the International Commission of Jurists, was founded in Britain in 1957 and currently has about 1500 members. Its aims are 'to uphold and strengthen the principles of the rule of law, to assist in the maintenance of the highest standards of administration of justice, and in the preservation of the fundamental liberties of the individual'. Justice's chief concerns about the THORP development were centred around the possibility that the UK is about to take a first, but important and possibly irreversible step leading to the modification or loss of its free institutions and fundamental liberties. Specifically, it argued that the following points were of particular concern: (i) the creation of expansion of an armed force, not necessarily forming part either of the armed forces of the crown, or of the ordinary police, (under the Atomic Energy Authority (Special Constables) Act 1976); (ii) greatly expanded surveillance by means which in the past the population has always found distasteful; (iii) the drawing up of plans, and the necessary legal power to carry them out, for dealing with any credible threat to public safety. It concluded that these and other potential threats to civil liberties seemed a very high price to pay for an attempt to close a possible future energy gap.

National Council for Civil Liberties

The National Council for Civil Liberties was founded in 1934, currently has about 5000 members, and is experienced in developing the defence of individual freedom. The Council's objection to the development was that the increase in the amount of plutonium dispatched by road and sea to overseas clients can only be made safe by massive degrees of secret police surveillance, not only of the workers in the plant, and the men employed by the transport contractors, but members of the general public, political organisers, trade union officers, liberal organisations, radical organisations, all of whom will inevitably be monitored, especially those groups concerned with the environment. The Council was also concerned about the consequences of an increasing nuclear commitment, arguing that it would create a degree of social divisiveness on a scale greater than was seen during the CND campaigns of the 1960s.

National Peace Council

The National Peace Council was founded in 1908 with the objective of achieving the abolition of war. There are currently 79 organisations affiliated to it with a nominal membership of over six million. The NPC associated itself with the objections of other groups, but particularly pointed to: (i) the question of the proliferation of nuclear weapons; (ii) the problems of the disposal of radioactive waste; and (iii) the ethical question of whether people who face risks (from accidents) should take that choice themselves, or have it taken for them. The Council also called for massive research and development of alternative energy sources, pointed to the problem of public information on the nuclear debate, and questioned whether there was a proven need for nuclear energy.

Network for Nuclear Concern

The Network for Nuclear Concern arose specifically in opposition to BNFL's expansion plans, and includes a number of environment groups in the North-West, these being:
Half life Lancaster
Friends of the Earth, West Cumbria
Half life Furness
Half life West Cumbria
Friends of the Earth, South Lakeland
Friends of the Earth, Lancaster University
Ambleside Nuclear Concern

Concentrating its arguments on the areas of public health and safety the Network argued: (i) that there are major uncertainties that might be intrinsically unresolvable; (ii) that the institutional controls are inadequate for the task, and (iii) that major technical and political uncertainties remain regarding waste disposal, and that insufficient research into alternatives had been carried out.

A major part of the NNC's case concerned a critique of institutional arrangements for decision-making in those areas, e.g. health risks to the public from discharges, where closed expert groups excluded effective public participation.

Political Ecology Research Group

The Political Ecology Research Group, based in Oxford, was founded in 1976 and is an association of scientists and research workers, concentrating on the study of the interface between politics and the environment. The group was particularly concerned about the level of public information about technological development in general, and nuclear developments in particular. It cited its reasons for opposing the THORP development as: (i) because of the scant information made available to the public for a reasonable assessment of the risks and benefits of the nuclear power programme, and THORP in particular;

(ii) because in the opinion of the group profound doubts and deep divisions exist within the scientific community on the issues of environmental health and safety, and on the social, political and ecological consequences of a nuclear-led increase in energy consumption. With regard to the safety of the public the group was particularly concerned with hazard assessment of major accidents and environmental health considerations arising from liquid and gaseous discharges, particularly to the marine terrestrial ecosystems.

Scottish Campaign to Resist the Atomic Menace

SCRAM is an umbrella organisation of Conservation Society and Friends of the Earth branches in Scotland. Its objection concentrated on three points: (i) that the levels of pollution in the Irish Sea, the Solway Firth and surrounding areas are already high by American standards as a result of the discharge of low-level wastes from Windscale, and that the proposed expansion would increase these pollution levels; (ii) that it is irresponsible to continue with, and seek to expand a process before methods of waste disposal have been devised, tested and approved; and (iii) that there is cause for concern that material transported to and from Windscale could, in spite of stringent security, get into the wrong hands.

Socialist Environment and Resources Association

The Socialist Environment and Resources Association was founded in 1973 with the aim of bringing environmental concern to the attention of the Labour movement, arguing that it is the workers, and poorer sections of the community who suffer most from environmental degredation. It has over 500 individual members including more than a dozen Labour MPs, and over 50 trade unions, co-ops and Labour party constituency branches are affiliates. SERA opposed the development on three basic grounds: (i) that employment in the planned instalation would be of a qualitatively different nature to employment in other power generating and processing industries, arguing that there is extensive and systematic curtailment of employment and trade union rights in the industry; (ii) that doubts could be thrown on the ability of workers adequately to operate the various provisions of the Health and Safety at Work Act; and (iii) that THORP was an extremely expensive way of creating a relatively small number of jobs, and that of those created half at most would be relevant to the locality because of the sophisticated nature of the job skills required.

Society for Environmental Improvement Ltd.

The Society for Environmental Improvement was founded in 1973 with aims as expressed in its title. It has a nominal membership only and attempts to achieve its objectives chiefly through the influence of its supporters. The Society confined its role at the Inquiry to presenting techniques which offer with varying degrees of certainty, an alternative to nuclear power, arguing that there was considerable public hostility to nuclear power and that consideration of alternatives was vitally relevant. Other sources of energy that could replace nuclear power that the society considered included: coal, solar power, wind power, hydro-electric power, photo-voltaic cells, tidal power, heat pumps, bio-fuels, geothermal power, marine sediment, pumped storage schemes, fuel cells, oil and gas, and the extension of conservation techniques.

Society of Friends, Keswick

The Keswick congregation of the Society of Friends represented a significant body of Quaker opinion throughout the country. It based its objections on a moral responsibility to future generations, and on the ethical aspects of effects which it was reasonable to believe would stem from the proposed expansion. It was particularly concerned about: (i) the disposal of high level wastes, and the safety of their storage; (ii) the dangers accompanying the transport of plutonium and spent fuel; (iii) the police and military surveillance which would become necessary if plutonium were to be stored permanently in the UK, or awaiting shipment abroad; and (iv) the levels of radiation from the discharges of so-called low-level wastes into the sea and atmosphere.

Town and Country Planning Association

The Town and Country Planning Association was founded in 1899 by Ebenezer Howard as the Garden Cities Association, and claims to be the oldest organisation in the world concerning itself planning and environmental matters. The Association advocates and promotes an understanding of national regional and urban planning policies, that will improve living and working conditions, safeguard the best countryside and farmland, enhance natural, architectural and cultural amenities and advance economic efficiency, so administered as to leave the maximum freedom to private and local initiative consistent with those aims. The Association had 12 grounds for objection to the THORP proposals which were: (i) That it has grave potential hazards in it which would constitute a very serious threat to the natural environment. (ii) The need for, and the purpose of the particular proposals as so far presented are not substantiated by the materials so far made available. (iii) The projected nature and extent of the so-called energy gap, and the present ill-defined energy policy are both seriously open to doubt. (iv) A serious questioning of the economic advantage to the nation of investment in the proposed development. (v) The uncertainty as to how many of the important radioactive wastes can be safely disposed of, there are also potentially grave risks involved in their containment. (vi) The risk of radioactive injury and contamination which may arise both for local people, and those over a much wider area are likely to be substantially greater than has so far been understood or anticipated. (vii) Employment more suitable for the present and future needs of the area could plainly be provided with the capital involved in the project. (viii) Very substantial improvements should be made in arrangements for transporting spent fuel. (xi) There is a serious cause for concern about the opportunities provided by this proposal for an increase in proliferation of materials capable of being used for military purposes by other nations for terrorism or sabotage. (x) The proposal gives rise for concern about the implications for society in the light of the inevitable security arrangements (xi) A thorough and detailed environmental impact analysis should be carried out before any decision is made. (xii) The supply of energy from plutonium is not indispensable to our society, though in its absence we might need to deal with some other forms of energy supply. Encapsulating the Association's position in a phrase, Robin Barratt, its junior counsel, in his closing speech characterised it as 'not yet, if at all'.

Windscale Appeal

The Windscale Appeal was an *ad hoc* grouping of ecological and conservationist organisations which came together solely for the purposes of the Inquiry to present a common front with counsel, expert witnesses, etc. Its constituent groups were:
- Concerns against Nuclear Technology Organisation (CANTO)
- Conservation Society
- Cornwall Nuclear Alarm
- The Ecologist Magazine
- The Ecology Party
- Greenpeace (London)
- Irish Conservation Society
- Society for Environmental Improvement Ltd.
- Wexford Nuclear Safety Association.

The grounds for the Windscale Appeal's opposition were: (i) That the establishment of the plant would represent a significant and irreversible step in the use and availability of plutonium in this and other countries with the consequent likelihood that it would sooner or later fall into the wrong hands, and/or be put to non-peaceful purposes, and that in any event security measures would be required that are not consistent with the maintenance of a free and democratic society. (ii) The plant itself is likely to prove unacceptably dangerous. (iii) The reprocessing of spent nuclear fuel results in substantial quantities of deadly waste for which no acceptable means of disposal now exist. (iv) The development would be a misallocation of national resources. In their closing submission the Windscale Appeal also drew attention to a number of procedural points of concern, related to the inquiry itself. These were: (i) the unsatisfactory nature of the adverserial approach at the inquiry; (ii) the disproportionate resources of the parties, BNFL having access to public funds, while the objectors did not; (iii) the fact that security aspects related to the inquiry could not be investigated; and (iv) the short length of the notice given of the commencement of the inquiry, leaving little time for preparation, and the rapid pace at which it proceeded with sittings on five days per week.

Windscale Inquiry Equal Rights Committee

The Windscale Inquiry Equal Rights Committee was formed after the announcement of the Public Inquiry with the aim of ensuring the widest possible discussion of the issues surrounding the Inquiry. Their submission that the development be delayed was based on six grounds which were: (i) Until the technological problems that BNFL are still wrestling with can be examined in greater detail; (ii) the long-term problems, both physical and political, of storage can be examined. (iii) the security implications of the THORP commitment can be examined by a separate body; (iv) public attitudes to tolerable risk can be examined and note taken of them; (v) alternative energy scenarios and resources can be properly researched; (vi) the role of the present nuclear institutions can be reappraised.

THE RESOURCES OF INQUIRY PROTAGONISTS

Comments have been made by a number of Inquiry participants during and since the Inquiry about the cost of maintaining a presence at it. We felt it useful to try in some way to quantify what financial commitment had been made by those involved. Mr. Justice Parker's report puts a figure of £145,000 as the cost of the Inquiry itself, while Professor Pearce's Windscale Assessment and Review Project (WARP) estimated the total cost involved at at least £1,250,000. With their greater financial resources, the largest proportion of this sum was obviously spent by those parties in favour of the proposed expansion. With the exceptions of the Uranium Institute and Copeland Borough Council they felt unable to give us figures of their expenditure at the Inquiry. Pearce et al. (1) give a figure of £750,000 as being the approximate sum spent by BNFL during the Inquiry. This would account for the lion's share of the 'pro-expension' side as the SSEB and the CEGB kept relatively low profiles throughout the Inquiry. Copeland Borough Council estimated their costs at about £7500, but of this only about £100 were 'extra costs' in the sense that those who gave their time to the Inquiry would have been employed by the Council in any case. The Uranium Institute estimated a total expenditure of £500.

Of those who opposed the application by BNFL, the great majority replied to our enquiries, sometimes in great detail. Figures cannot be *precise* as not all groups replied, and others were unable to disentangle costs of one sort or another from their total costs. However, some useful indications can be given. Total expenditure by objector groups at the Inquiry was between £150,000 and £160,000. £70,000 was spent by the Government of the Isle of Man alone. A total of just under £100,000 went in legal fees, £50,000 of this by the Isle of Man Government. Other approximate costs were £15,000 for administration, £8000 for travel costs, £7000 for witness fees (£5000 of which was by the Isle of Man Government), £5000 for accommodation, £3500 for research, and a further £15,000 that was not disaggregated. When asked what they felt they would have needed to spend in order to present their case fully and without individual financial hardship, objector groups as a whole estimated that between £300,000 and £325,000 would have been sufficient, i.e. roughly a doubling of their actual expenditure. This figure was broken down thus: legal fees: £175,000; administration: £24,000; witness fees and expenses: £22,500; research: £20,000; accommodation: £9000, general: £40,000. These figures were calculated on a basis of limited public funding being available. The most remarkable figure is that proportion which would be devoted to research, a six-fold increase. It also supposes a continued major presence of the legal profession, but probably underestimates costs related to witnesses, travel and accommodation, as a number of respondents put comments to the effect that such categories were not applicable as they could not afford witnesses who charged fees. This figure of about £300,000 is still only around one-third of the total sum expended by those in favour of the development. In the event the disparity was about 6:1 in favour of the development. The question of public funding for objectors at future major public inquiries has been raised in a number of publications since the inquiry (2).

Sources of funding for those in favour of the Inquiry were general funds for the local councils, electricity boards and BNFL, i.e. the ratepayers in the case of the councils, the British taxpayer in the case of BNFL, and the electricity consumer in the case of the two boards. The sources of finance for objectors were more varied. A large proportion came from public subscription, though the Friends of the Earth organised Windscale Fighting Fund, and the Windscale Appeal's Campaign of the same name. A lesser, but nevertheless significant sum was channelled to a wide variety of objector groups through the so-called 'Goldsmith Fund', a fund-raising exercise organised and co-ordinated by Sir James Goldsmith, brother of Edward Goldsmith the editor of *The Ecologist*. Somewhere in the region of £20,000 came from this source. The remaining finance was taken from the general funds of individual groups, with the exception of course of the £70,000 spent by the Isle of Man Government, which would have come from the general funds of the government, raised by taxation.

The total expenditure by the various groups at the Windscale Inquiry is of relevance in relation to any future public inquiries into nuclear matters, and the proposals currently under consideration for the public funding of objector groups. With more care about avoiding duplication of effort on the part of objectors, and a reasonable fee-scale for legal representatives we conclude that a total of £250,000 at 1977 prices would be the kind of figure necessary.

A BRIEF ANALYSIS OF RELATIONSHIPS BETWEEN THE PARTICIPANT GROUPS

1. Those opposed to the development

The presentation of the opposition case at the Windscale Inquiry was notable for the number of groups registered as opposing the development, and for the considerable extent to which their cases overlapped and repeated each other. Attempts were made before the start of the Inquiry to present a united front, but disagreements over policies and tactics, and personality clashes meant that co-operation between groups at the Inquiry was largely on an informal basis only. One organisation, the Windscale Appeal, came into existence solely for the purpose of appearing at the Inquiry, and has since disbanded. This body consisted generally of smaller groups who would have been unable individually to present a case to the inquiry. (See Windscale Appeal pp.32–33 for its constituent members.) The counsel for this group was David Widdicombe QC. He was also counsel for Justice, and the Council for Science and Society. The Society for Environmental Improvement while remaining a member of the Windscale Appeal for financial purposes presented its own evidence. SEI also represented the National Centre for Alternative Technology which was not itself a member of the Windscale Appeal. The only other umbrella organisation at the Inquiry was the Network for Nuclear Concern, which brought together a number of conservation and anti-nuclear groups in the north-west.

Of NNC's members the only group to present a separate case was Friends of the Earth, West Cumbria.

The Windscale Appeal developed out of a series of meetings in London immediately after the announcement of the Inquiry. It had originally been intended that objector groups should present a common front at the Inquiry, but it soon became clear that differences of philosophy and tactics would make this impossible. The result was the creation of the Windscale Appeal, and a more loosely grouped series of objectors around the coordinating centre of Friends of the Earth, and its fighting fund. This latter group included SERA, NCCL, the Natural Resources Defence Council, and at a slightly greater distance NNC, PERG, and the TCPA, though of these only SERA benefited financially from the fighting fund. Although overlapping of cases amongst these groups did occur, attempts were made to keep it to a minimum through this informal coordinating structure; thus FOE concentrated on economic factors, arguing for a ten-year moratorium on any development, while SERA concentrated on employment and industrial relations aspects, NCCL on civil liberties, and NNC, PERG and TCPA on environmental health and safety.

These two basic sets of objectors were held together financially by their respective public appeals, the Windscale Appeal for the group of that name, and the Windscale Fighting Fund for the others. Both groups were also beneficiaries of the 'Goldsmith Fund', which raised money for the objectors from institutions in the City. Windscale Appeal and its constituent groups, Friends of the Earth, TCPA, PERG and the NNC all benefited to varying degrees from this source of funds, but Sir James Goldsmith excluded the NCCL, on the grounds that it was 'political'. The Windscale Appeal did not suffer a similar exclusion in spite of the Ecology Party being a constituent member.

At the Inquiry itself the stresses and strains between the various objector groups, particularly those between the Windscale Appeal and the rest, were largely submerged in concentrating on a common objective. As the Inquiry progressed, cross-fertilisation of ideas between the groups grew, and assistance on cases was frequently given. Windscale Appeal and SCRAM shared a witness in Professor Tolstoy, as did Friends of the Earth and the Society for Environmental Improvement in Gerald Leach. Cooperation between experts, advocates, and witnesses was particularly marked between PERG, NNC and the TCPA; FOE and NCCL (along with the CPRE who did not appear at the Inquiry) had jointly commissioned a pamphlet entitled 'Nuclear Prospects', on the potential infringements of civil liberties that an expanded nuclear power programme could present, and one of the authors appeared as a witness for Justice. David Widdicombe, QC, counsel for the Windscale Appeal also appeared for two other objector groups, Justice, and the Council for Science and Society, while Oliver Thorold, junior counsel for Friends of the Earth also acted on behalf of the Natural Resources Defence Council. One of the

NRDC's staff also appeared as a witness for the TCPA. In general cooperation between objector groups at the Inquiry was much as one might have expected; a certain amount of mutual suspicion and bitterness at the commencement of the proceedings mainly due to the collapse of the proposed common front. However, a gradual drawing together in mutual assistance and cooperation occurred as the inquiry progressed, in spite of philosophical and ideological differences as the realisation that they were all engaged in a common cause became clearer. Suggestions that the level of cooperation and dual or multi-membership by individuals of various groups was evidence of a Kremlin inspired plot were put forward at the Inquiry (3) and in a recent book (4) but seem far-fetched. Indeed Mr Justice Parker firmly dismissed them in his report (para. 7.25.)

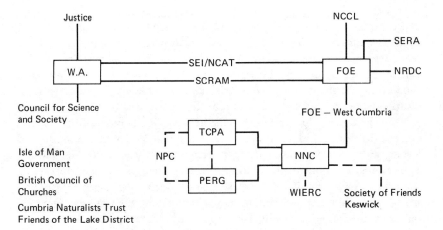

Solid line indicates formal association
Broken line indicates informal association
Name enclosed in box indicates beneficiary of 'Goldsmith Fund'

Notes: 1. SERA were given money from Friends of the Earth Fighting Fund.
2. NNC and TCPA were associated with each other's cases on health and safety matters, TCPA originally became involved in the controversy when 'Half-Life' approached their planning aid service for planning assistance.

Figure 1. Formal and informal relations between objector groups.

2. Those in favour of the application

Of the five main parties in favour of the development proceeding at Windscale Cumbria County Council and Copeland Borough Council are the relevant statutory local authorities. The other three, British Nuclear Fuels Ltd., the Central Electricity Generating Board, and the South of Scotland Electricity Board, are all publicly owned authorities although provision was made when BNFL was set up (under the Atomic Energy Act 1971) for a proportion of the share capital to be privately subscribed. In the event none has been, and the United Kingdom Atomic Energy Authority holds all the share capital in trust for the Secretary of State for Energy. In that BNFL is involved in all aspects of the nuclear fuel cycle, and nuclear power in Britain is used to generate electricity, the links between BNFL and the electricity boards must, perforce, be quite close.

BNFL has a number of associated and subsidiary companies, these include: BNFL Enrichment Ltd., a company wholly owned by BNFL and the Central Subsidiary through which all financial aspects of participation in the tripartite centrifuge project are handled. Among the associated companies are: (i) Centec GmbH, a company incorporated in Germany in 1971 for purposes connected with the development, design and manufacture of centrifuges and plants for the enrichment of uranium by the centrifuge process. BNFL holds 33% of the shares, the remainder being held by Germany and British shareholders. (ii) Urenco Ltd. incorporated in England in 1971 with the purpose of owning and operating, through a partnership established in the UK and the Netherlands, enrichment plants using the centrifuge process and marketing enriched uranium produced in these plants. Again BNFL has 33% of the shares, with Germans and Dutch shareholders taking the rest (see figure 2). (iii) United Reprocessors GmbH incorporated in Germany in 1971 for the purpose of marketing and providing services for the reprocessing of irradiated fuel from nuclear power stations using uranium dioxide fuel. BNFL has 33% shareholding, as do Cogema (France) and KEWA (Germany).

BNFL is also involved in (a) Nuclear Transport Ltd., having a 33% share, as do French and German participants; (b) Pacific Nuclear Transport Ltd., having a 75% share, with Japanese interests holding the remaining 25%; (c) Combustibli Nucleari SpA, BNFL 50%, Italian interests 50%; and (d) Nukleardienst GmbH, BNFL 50% shareholding, German interests 50% shareholding.

On the European scene Nukem, itself one of the participants, through URANIT in Urenco (see below) is owned by RWE (Rheinisch-Westfalisches Elektriziatswerk Aktiengesellschaft) of Essen (45%). Degussa of Frankfurt (35%), Metallgesellschaft AG of Frankfurt (10%) and the Imperial Smelting Corporation Ltd. (a subsidiary of Rio Tinto Zinc, London) (10%). RWE is

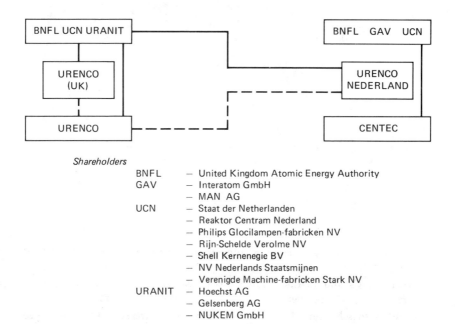

Figure 2. The organisation of Urenco/Centec. (From: *Nuclear Engineering International*, December 1977.)

also the major shareholder in SBK (Schneller-Brüter Kernkraftwerkgesellschaft) with 69%, SEP NV (Samenwerkende Elektricitiete-Produktie Bedrijven) having 14.75%. Synatom (Synatom, Société anonyme, Bruxelles) and the CEGB (Central Electricity Generating Board, London) 1.6%. SBK is the company formed for the construction of the Kalkar (SNR 3000) FBR. SBK also holds a 16% share in NERSA (Centrale Nucleaire Européenne à Neutrons Rapides SA) the company which owns the Creys-Malville FBR in South West France, the other shareholders in NERSA being EdF (Electricité de France) 51% and ENEL (Ente Nazaionale per d'Energia Electrica) Italy 33%. Thus the CEGB is involved directly in the construction of the Kalkar FBR and indirectly in SBK in the Creys-Malville FBR; it is also associated through its shareholding in SBK with RWE, which itself is associated with BNFL, through its shareholding in URANIT, one of BNFL's partners in URENCO.

The relationships in Britain are more straightforward as the electricity boards are the only customers for nuclear power stations in Britain, and buy electricity from the UKAEA Calder Hall power stations, and the 250 MW(e) Dounreay DFR. The nuclear power industry in Britain seems to be in a state of permanent crisis but the current commercial structure involves BNFL (through the UKAEA) quite closely in the National Nuclear Corporation, the

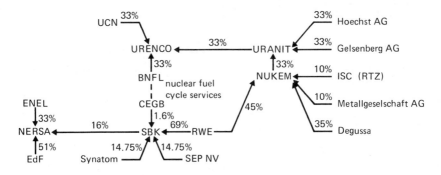

Figure 3. The BNFL/CEGB Relationship in Europe.

only British reactor construction company presently in existence. The CEGB and SSEB are the NNC's only customers and will shortly place orders for further AGR's and in due course for a PWR; but before this happens, a reorganisation of the NNC seems likely, in which the UKAEA is likely to reduce its shareholding, BNFL may take a direct interest, and Rolls-Royce and Vickers may be brought in as well.

Notes

1. Pearce, D., Beuret, G. and Edwards, L. (1979) *Decision-making for Energy Futures: A Case Study of the Windscale Inquiry.*
2. For example, TCPA (1979) Financial assistance for objectors at major public enquiries; PERG (1978) Public participation and energy policy; Pearce et al. (1979) *op. cit.*
3. For example, Transcript, day 97, pp. 25–63.
4. Hoyle, F. (1977) *Energy or Extinction* (London:Heinneman).

Index to the Issues

1. **World and UK Energy Demand**

    1.1. Projections of energy demand
    Official forecasts
    General critique of forecasting
    Alternative forecasts

    1.2. Nuclear Strategies
    Uranium supply and conservation
    UK and EEC nuclear industry implications
    Projected commitments
    Strategic, financial and trade implications

    1.3. Alternative Non-nuclear Strategies
    Assumptions of alternatives
    Risk assessment

    1.4. Summary of Assumptions and the Case for Nuclear Strategies

2. **Weapons Proliferation**

    2.1 International Agreements
    Non Proliferation Treaty
    Present and potential nuclear powers

    2.2 Plutonium
    risks and safeguards
    provision of reprocessing facilities

    2.3 Alternative Fuel Cycles

3. **Security and Civil Liberties**

    3.1. Security and the Control of Technology Policy
    Accident analysis
    Discharge control

3.2 Security and the Rights of the Workforce
   Safety and strike action

3.3 Security and the Problem of Terrorism or Sabotage

3.4 Civil Liberties and Nuclear Expansion
   THORP
   The plutonium economy

4. Health and Safety

4.1. Comparative Aspects of the Reprocessing Option
   Summary of arguments on health and safety

4.2. The Basis of Control

4.3. Background to Risk Assessment
   History of discharges to the environment
   Storage and disposal
   Past radiation exposure of the workforce
   Past radiation exposure of the public
   Risks to the workforce and to the public in the event of accidents
   Societal risks

4.4. Statements of Intent
   Waste disposal and discharges to the environment
   Solid waste disposal
   Radiation exposure of the public and the workforce

4.5. Institutional Safeguards
   The authorising bodies and radiation standards
   (i) the health of the workforce
   (ii) the health of the public
      genetic risk
      somatic risk
      models of the biological behaviour of radionuclides
      the acceptability of exposure standards
   The competence of the authorising bodies
   (i) protection of the workforce
      health statistics
      exposure standards
      general housekeeping
   (ii) protection of the public
      whole body monitoring
      strategies of environmental monitoring

        Ravenglass
        waste management strategies
        alternative waste management programmes
        site selection
   (iii) research and organisation
   Accountability and acceptance of control
   Intentions

5. Conventional Planning Issues

   5.1 Suitability of the site in relation to the national park

   52. Amenity

   5.3 Infrastructure

   5.4 Local employment and training schemes

   5.5 Economic effects

6. Democratic Accountability

   6.1 Participation and Accountability
       Policy
       Standards and control

   6.2 The Inquiry Procedure
       Environmental impact statements
       Financial disparity

How to use this Guide

Each broad issue has been accorded a section, e.g. *Health and Safety*. Within each section the issue is further broken down and particularly contentious issues are examined on a case by case basis, e.g.

Issue: Whether the authorising bodies were sufficiently open to a revision of standards.

The next outlines the evidence and argument and is comprehensively cross-referenced to the Inquiry Transcript, e.g.

'. . . as relied upon by BNFL' (Schofield 21.84E–87F).

The name of the witness is given, followed by the number of the daily transcript, in the above example: day 21, and then the page references, page 84, starting at paragraph E through to page 87 finishing at paragraph F. The daily transcript itself carries an index of names for each day's proceedings, so that given the day and the name, the full evidence of that person may be found. In almost all cases, the issues run over several days and may be dispersed between other issues and other witnesses. Thus at the end of each particularly contentious issue we have provided a summary of case references.

In addition to the names and the reference to the transcript, we have also made reference to those documents of particular importance to that case, e.g. (BNFL 187, PERG 35). The document number is that given at the Inquiry by the Department of the Environment. A full list of all these documents appears as an annex to Volume 2 of the Parker Report, and we have not therefore repeated this list. We have assumed that all researchers will have access to the full Parker Report. Volume 2 also contains a list of all the participants with their affiliations and qualifications.

1. World and UK Energy Demand

1.1. PROJECTIONS OF ENERGY DEMAND: OFFICIAL FORECASTS

Issue: How official UK forecasts are arrived at, and the criteria applied in formulating them.

1.1.1. Jones in his statement (42. 50A–52F) outlined the Department of Energy's general policy, objectives etc.:– 'to secure that the nation's energy needs are met at the lowest cost in real resources consistently with achieving adequate security and continuity of supply, and consistently with social, environmental and other policy objectives'. Under cross-examination by Kidwell (42. 83C–99E) he outlined the Department's forecasting criteria: (i) the level of economic growth; and (ii) world energy prices, there being a 1.7% primary fuel consumption increase associated with a 2.7% pa GDP increase. He further stated that the Department's thinking on an energy conservation component included an allowance of 20% for energy conservation.

1.1.2. Silsoe drew attention to G 3 and BNFL 9 on the likelihood of energy becoming scarcer and dearer over the next 30 years as oil and gas supplies level off, and in due course, decline, while the world demand for energy continues to rise. He argued that the UK energy policy as set out in G 3 was directed to ensuring that the UK needs were met at minimum real cost consistent with achieving security of supply while meeting social, environmental and other policy objectives (i.e. repeats Jones' statement). The criteria used when considering UK energy demand and supply projections (1975–2000) were: (i) energy demand based on economic growth of 2.3% pa with energy prices either remaining constant or doubling in real terms by the end of the century; (ii) energy supply based on varying assumptions about resources and technical potential. The mix of supply will depend on the relative cost of each fuel.

1.1.3. Wright (40. 6F–H) drew attention to the drop in energy demand since the 1973 crisis, in relation to official forecasts for energy demand, and stated that the large projected capacity of THORP would now serve both UK and foreign customers.

1.1.4. Barratt (40. 83B−86A) drew attention to CEGB 1, which forecast (p.19) an electricity growth rate of 3.5% pa from 1975/76 to 1995, leading to a maximum demand in 1995 of 81 GW, with an alternative case where growth of demand was only 1.5% pa to 1995. Barratt argued that if the latter forecast was the more accurate there would be no requirement for an additional nuclear capacity. Wright replied that additional capacity of some sort would be needed after 1995, and therefore additional plants would need to be started by 1990.

1.1.5 Kidwell in cross-examination of Allday (6. 22F−26G and 8. 23F−34G) cited FOE 14, *The Myth of Uranium Scarcity* as evidence for the FOE case that there was likely to be sufficient uranium supplies into the 21st century, at prices that would make plutonium fuels and reprocessing uneconomic.

1.1.6. Parker (para. 8.37) comments, 'with so many uncertainties the only prudent course is to adopt a strategy which will give the greatest assurance that no matter how the variables change, the energy needed to support an acceptable society can be provided'.

Case references
Allday 6.22F−26G and 8.23F−34G; BNFL 2.8C−26E; FOE 42.97E−H; Jones 42.83C−99E; TCPA 40.83A−86A; Wright 40.6F−H and 83B−86A.

Proofs
Wright para. 10

Documents
BNFL 9; CEGB 1; FOE 14; G 3

PROJECTIONS OF ENERGY DEMAND: GENERAL CRITIQUE OF FORECASTING

Issue: The veracity or otherwise of the Department of Energy's energy forecasts.

1.1.7. Kidwell drew attention to Leach's proof, (especially para. 16) in which the Department of Energy's forecasts are criticised for giving undue emphasis to previous historical experience.

1.1.8. Leach pointed to the rapid downward shift in the Department of Energy's projections of future energy demand and to the considerable increase in attention paid to renewable energy systems in successive energy policy documents. He criticised the underlying assumptions of the Department's energy projections; economic growth, varying between 2 and 3% pa with energy prices remaining about constant or doubling in real terms, by the end of the century. He argued that a doubling in price was the minimum to be expected. In the section of his proof entitled 'Critique of the Department of Energy's demand forecasts', he argued that the forecasts have a built-in tendencies to over-estimate future energy demand, and hence the need for nuclear power; and that the forecasts rely heavily on two cardinal assumptions, both of which he challenged. They were: (i) that energy growth must continue to be closely tied to economic growth; and (ii) that future growth trends will be roughly exponential in nature since they have been in the past. He then continued by citing sections of G 3 to support this statement.

1.1.9. He also criticised the two part breakdown of industrial energy consumption ('iron and steel' and 'other') as being too crude. The Department justified this approach on the grounds that they cannot forecast the future 'mix' of industry and therefore can see no point in a further breakdown. Leach went on to argue against the use of exponential growth curves in forecasting, arguing that in future they should be either linear or 'saturation' curves.

1.1.10. Chapman argued that there was no such thing as an 'energy gap', references to an 'energy gap' usually referring to a situation in which the production of energy from indigenous sources is less than indigenous demand, a situation that has pertained in the UK for the past 20 years. He argued that projections of an energy gap by 2000 were based on an unrealistically high demand forecast and a pessimistic supply forecast. He further argued that a situation in which the supply of energy was less than the potential demand leading to price rises caused (a) further exploration; (b) new energy sources becoming economic; (c) the encouragement of more efficient use of fuel and energy conservation; and (d) the discouragement of marginal uses of fuel. He suggested that the 1973/74 energy crisis was in large measure due to having such a great dependence on coal and oil. He continued his proof with a critique of official forecasts of energy reserves (e.g. gas) and pointed to contradictions between the Department of Energy's estimates of reserves and the British Gas Corporation's. On the basis of these data Chapman argued that it was likely that indigenous supplies of conventionial fuels could satisfy UK demand beyond 2000.

1.1.11. Davoll criticised assumptions of energy demand made on the basis of recent past experience, a period of cheap energy and rapid material growth, and argued that technology would be better directed towards the achievement of elegance and parsimony in managing demand rather than towards the blind expansion of supply.

1.1.12. Odell, criticised the Department of Energy (especially their Energy Policy Review Paper) for leaving many questions unasked, while giving answers to others based on a limited appreciation of the alternative options open to the country. The Department, he argued, made no real effort to attach reasonable probabilities to important energy supply alternatives such as development of indigenous supplies of oil and gas. Furthermore they were assuming rates of development of energy demand over the next twenty five years, a period in which energy prices are likely to increase in real terms, which would be higher than any other period of 25 years in modern British economic history.

1.1.13. Page challenged the recent forecast prices of nuclear fission generated electricity.

1.1.14. Parker (para. 8.38) commented 'some parties and witnesses appeared completely to overlook the fact that there is a great difference between ... making a confident forecast, without having either power to act upon it, or any responsibility for the consequences, if when someone else has acted upon it it proves to be wrong and ... taking, and acting upon a decision the consequences of which will affect the lives of millions. It is the government which has the power and duty to make such decisions'.

1.1.15. He further outlined (paras. 8.39–40) why he considered that, in spite of the uncertainty of forecasts, the government should proceed with the development of nuclear power.

Case references
Chapman 56.39H–44E and 57F–60A; Davoll 61.59F–69D; FOE 42.86E–99E; Leach 54.23A–37H; Odell 77.78H–85B; Page 35.9H–24D

Proofs
Chapman paras. 1–16 and 35–43; Davoll paras. 2.1–8.4; Leach paras. 15–71; Odell paras. 4.1–4.7

Statement
Jones

Documents
BNFL 198; G 3

PROJECTIONS OF ENERGY DEMAND: ALTERNATIVE FORECASTS

Issue: Whether BNFL's forecasts of energy demand, and the subsequent assumed need for THORP, were correct or not.

1.1.16. Blackith criticised the common assumptions made about continued economic growth, and therefore of energy consumption, and argued that a steady-state economy was a preferable alternative.

1.1.17. Odell outlined an alternative energy scenario which included a modest increase in coal production to 135 million tons per annum, a 2 mtce contribution by non-nuclear renewable resources, small increases in hydro and nuclear power, as commissioned stations came on stream, with the remainder of the energy demand to be met by oil and natural gas. He concluded by arguing that no self-evident energy gap could be seen to arise in the UK before 2000, and that any decision related to a nuclear alternative could be postponed to 1980—83.

1.1.18. Chapman argued that the recent historical trends in fuel consumption were unlikely to persist. He cited growth in home heating and personal transport as the major factors in the recent past, factors which could not be expected to continue.

1.1.19. Leach (54.38A—51D and 55.12C—27B) outlined the alternatives of insulation, fluidised bed combustion, and electric heat pumps, and their costs in comparison to nuclear power costs, and argued that all these methods of energy saving could usefully be used in existing housing. He also drew attention to the savings which could be made by solar water heating and solar photovoltaic devices. He argued that as the growth in the number of households slows and stops (as population growth slows, and family fission is reduced), growth trends of domestic fuel consumption will also slow, and that with proper insulation it is perfectly feasible to reduce primary fuel demand for domestic space and water heating to one-third of that of comparable new or post World War II housing. This accounts for 85% of energy consumption in the domestic sector, itself 25% of total national primary fuel consumption. The potential in commercial premises was also seen from various American and DOE Property Services Agency studies, to be considerable. He also outlined the conservation possibilities in the industry and transport sectors.

1.1.20. Silsoe in cross-examination of Leach (55.28A—84F) argued that energy saving options did not invalidate the need for THORP, and also pointed to some of the problems of massive insulation and conservation programmes; CHP schemes not being economic; and the problems of condensation in some houses. He also suggested that of the home insulation programmes outlined in Leach's proof (table 5, p. 20) only the first four steps were marginally cost effective.

1.1.21. Parker (para. 8.37) remarked that the evidence from alternative forecasts 'fell far short of what I would require were it for me to make a definitive forecast'.

Case references
BNFL 55.28A—84F; Chapman 56.44F—49G and 55H—57E; Leach 54.38A—51D, 55.12C—27B and 28A—84F; Odell 77.85C—86B

Proofs
Blackith paras. 5.1−5.10; Chapman paras. 17−34; Leach paras. 72−117 and 142−189; Odell paras. 5.1−5.9. (Note: the relevant section of Blackith's proof was not read into the transcript.)

Documents
BNFL 176, 235, 236; FOE 34, 105, 110, 111, 114

1.2. NUCLEAR STRATEGIES: URANIUM SUPPLY AND CONSERVATION

Issue: Whether, from the point of view of encouraging the conservation of energy sources, it was sensible to reprocess spent fuel in order to extract the plutonium for re-use, or whether this was both unnecessary and uneconomic because uranium was in cheap and plentiful supply, so making storage of spent fuel preferable.

1.2.1 Allday (3.43B−44A) stressed the importance of reprocessing as a contribution towards energy conservation, in that 'unburnt' uranium could be extracted and utilised, and the plutonium arising from the fission process, could also be utilised, either in thermal reactors, or FBR's. He claimed the energy savings by FBR's were some 50 times greater than in existing types of reactor, and argued that in economic and resource conservation terms there was a strong case to be made for reprocessing, in order to make plutonium available should the FBR option be pursued later.

1.2.2. In cross-examination by Widdicombe on his proof para. 31 (7.36C−48F), Allday admitted that BNFL had done a 'return on capital' study of the reprocessing plant. Widdicombe asked to see this, and Silsoe produced BNFL 174 and 175 which gave a breakdown of the expected costs of investment in THORP, and the estimated overseas income associated with THORP (7.45D−48F).

1.2.3. Kidwell, in cross-examination of Allday (6.22F−26G and 8.23F−34G) drew attention to FOE 14, *The Myth of Uranium Scarcity*, which argued that there was likely to be sufficient uranium supplies into the 21st century, at prices that would make plutonium fuels and reprocessing uneconomic. Allday disagreed with the documents conclusions, and did not accept the contention that current short-term shortages and high spot-market prices were the result of artificial demands created by the past nuclear fuel enrichment policies of the US government. Allday quoted paras. 26−27 of his proof to substantiate this view and stated that his main objections were to the pricing aspects of the document. As evidence against the FOE position Allday cited BNFL 41 (8.35A−D) which discussed likely future uranium prices taking inflation into account. Kidwell responded by citing the chapter in FOE 14 on the history or US uranium production and prices (8.43C–51G and 53B–79E) as substantiation for FOE's view that it was possible for the uranium industry to produce uranium profitably at steadily declining prices [due to improvements in exploration and production technology etc.] and further cited FOE 78, and 79.

1.2.4. Allday (8.73C–E) argued against this position by citing BNFL 41 which argues that constraints on future uranium mining are likely to be political and economic rather than geological. Kidwell replied by citing BNFL 177, p. 311 (8.77C–79E).

1.2.5. Price outlined the importance of uranium supplies in relation to the justification for developing nuclear reprocessing and argued that reasonable estimates of the growth in uranium requirements over the next 25 years would be around 10% per annum. He argued that it was necessary to consider the worldwide context, and not just the UK, as this would determine the pressure on uranium sources. He further postulated that world uranium demand would begin to run up against supply problems by 2010, threatening the future of nuclear power, unless methods were found to stretch these resources, by (a) reprocessing (15% gain); (b) reprocessing and breeding (which would transform the picture by the second half of the 21st century); (c) vigorous exploration. He concluded that he found it difficult to believe that uranium discoveries could continue to be made to match the rapidly growing period of demand, and so considered the long-term future of nuclear technology would be reliant upon the development of reprocessing and breeder technologies, and that optimistic scenarios based on an assumed abundance of natural uranium were not realistic. Kidwell in cross-examination of Price (52.32A–47E) argued that in fact since the oil crisis with the economic downturn, the long-term prices for uranium had not risen as expected, and that nuclear capacity at the end of the century would be lower than previously forecast, so reducing the demand for uranium.

1.2.6. Chapman (56.81G–83B) outlined the relative generating costs of an FBR and a Magnox reactor. He argued that with an assumed price of $30 per lb for uranium for a Magnox 'burner' reactor, a breeder reactor, even with free plutonium, would not be able to compete economically.

1.2.7. Page contradicted FOE's view that uranium prices would be likely to remain low, pointing out that they had increased eight-fold in the past few years.

1.2.8. Chapman (56.87G–88C) argued that in order to mine sufficient uranium to keep an AGR in fuel for its lifetime it was likely that 50–60 million tonnes of uranium ore would need to be mined, a similar quantity of material to that required by a coal-fired power station, hence there would be no environmental advantage from the mining component.

1.2.9. Silsoe in his cross-examination of Chapman (57.37F—38F) argued that reprocessing would negate the need for the mining of such quantities of uranium, and therefore would not entail the high associated environmental costs. Chapman agreed.

1.2.10. In response to the FOE case that reprocessing was unnecessary, as outlined above, and that storage of spent nuclear fuel was the preferable option, Allday, under cross-examination by Kidwell (4.42D—43C) agreed that management of spent fuel without reprocessing for periods of ten years, and up to 20 years, probably did not present very much technical difficulty. He did however consider unreprocessed fuel to be as dangerous or as safe, as reprocessed fuel.

1.2.11. Kidwell (4.46F—47F and 5.2B—8F) drew attention to a Canadian report, FOE 67, which outlined the options open vis-à-vis reprocessing and concluded that for the Canadian nuclear programme (CANDU) reprocessing was at the moment less economic than storage. Allday countered this by pointing out that Canada, unlike Britain, was 'blessed' with large uranium reserves. Kidwell then referred to FOE. 68 an ERDA report (5.8G—12G) and argued that this too considered the storage option to be a feasible one. He further cited FOE 71 (5.28C—34H) as evidence that the breeder system would not improve uranium utilisation 50-fold as Allday had claimed.

1.2.12. Rotblat argued for the storage rather than reprocessing of spent fuel rods, on the grounds that this would have the advantages of eliminating the need to process huge quantities of radioactive material, and there would be no release of low-level wastes into the sea.

1.2.13. Flowers, stated that short-term work by the UKAEA into the possibilities of storage had led them to conclude that within a ten-year period of storage in demineralised water spent fuel was not subject to significant corrosion. However on the evidence he concluded that he could not make an estimate of the maximum satisfactory pond-storage period, though it was more than ten, and possibly more than 20 years. Under cross-examination by the Inspector, Flowers admitted that the research carried out at the request of the Inquiry was the first into the possible safe lengths of storage of spent fuel undertaken by the UKAEA.

1.2.14. Parker (paras. 9.6—9.7) comments that taking into consideration the costs of storage, and the value of uranium recovered through reprocessing it would seem that storage was economically disadvantageous in comparison with reprocessing.

Case references
Allday 3.43B—44A, 4.42D—43C and 46F—47F; BNFL 7.45D—48F and 57.37F—38F; Chapman 56.81G—83B and 87G—88C, 57.37F—38F; Flowers

86.58B—65C and 80D—F; FOE 4.46F—47F, 5.2B—8F, 8G—12G and 28C—34H, 6.22F—28G, 8.23F—34G, 43C—51G and 53B—79E, 52.33A—47E; Page 35.11C; Price 52.21E—30D and 33A—47E; Rotblat 74.82H—86B; WA 7.36C—48F

Proofs
Allday paras. 19—31; Chapman Appendix 2; Flowers; Page; Price; Rotblat paras. 25—32

Documents
BNFL 36, 41, 175, 177, 305; FOE 14, 67, 68, 71, 78, 79, 109; G 3; Trib 6

NUCLEAR STRATEGIES: UK AND EEC NUCLEAR INDUSTRY IMPLICATIONS (CAPACITY)

Issue: Whether the THORP development should be seen in a Britain only context, or whether its intended size should be seen as an indication that its reprocessing facilities would be used on a Europe-wide basis.

1.2.15. Coates argued that the Windscale development should be seen within the context of a European nuclear policy, with the role of the EEC Commission as a proselytiser for nuclear power being a factor to take into account. She further argued (97.63B—65C) that THORP was not needed for British arisings, but rather in order to fulfill the contracts which are being entered into by United Reprocessors for its three partners, and that pressure was being put on BNFL by United Reprocessors and the EEC Commission to make THORP larger than they really needed it to be.

1.2.16. Taylor in cross-examination of Mr. Allday questioned him as to whether the size of the Windscale development might not be related to financing, economic and planning problems, relating to the proposed German reprocessing plant, and that as there were legal limits to the amount of stored spent fuel that could be kept in ponds on reactor sites, the Germans were faced with a choice of either temporarily closing down reactors, or having some of their excess spent fuel reprocessed at Windscale.

1.2.17. Allday agreed that it was possible that some German reactors might have to close down at some stage in the future but was unable to put a date on this. He said that the decision to go for a 1200 tonnes pa capacity reprocessing plant at Windscale was not related to this, as if the Japanese contract was signed, as they hoped, there would not be much room for German fuel.

Parker did not comment on this argument.

Case references
Coates 97.63B—65C; Taylor 8.81G—87C

Proof
Coates paras. 1.1—3.33

Document
BNFL 196; FOE 56; G 3, 31

NUCLEAR STRATEGIES: PROJECTED COMMITMENTS

Issue: Whether or not the existing B.205 plant would be able to handle the projected throughput, so making THORP redundent.

1.2.18. Kidwell cross-examined Allday (4.6D—10F) on the capacity of the currently built B.205 separation plant, and Allday suggested a maximum throughput figure of 2000 tonnes per annum (a reduction from 2500 tonnes pa as had been previously suggested in 1971).

1.2.19. Kidwell argued (4.10G—11B) that there was currently substantial excess capacity in the B.205 plant. Allday disagreed but accepted that there was some excess, in the region of 600 tonnes pa. He further explained that the present situation with the B.205 plan was that due to the three-day week and a shutdown in 1974 totalling ten months approximately 1000 tonnes of unreprocessed magnox fuel remained as backlog to be treated, the current throughput being 1300 tonnes pa and the current electricity board arisings being in the region of 1300—1370 tonnes pa.

1.2.20. Kidwell referred to BNFL 134 in support of his argument that the figure for throughput was in the region of 2500 tonnes pa and that in conjunction with the B.204 Head End plant this when reopened would be adequate for current reprocessing needs.

1.2.21. Parker (para. 33) comments that to use B.204/205 for reprocessing would leave no reasonable margin for untoward occurrences; that as B.204 is old it must be less safe for its operators than a new plant, especially if worked to capacity; and that the B.204/205 reprocessing route would in any case not be capable of handling arisings from the additional reactors that seem likely to be ordered, or the 1150 tonnes of foreign fuel, the reprocessing of which BNFL are already committed to.

Case references
Allday 4.6D—10F and 10G—11E; FOE 4.18C—21H

Document
BNFL 134

NUCLEAR STRATEGIES: STRAGEGIC, FINANCIAL AND TRADE IMPLICATIONS (RESOURCE DIPLOMACY).

Issue: To what extent international financial and trading pressures had an influence on the economic viability of reprocessing and in which direction.

1.2.22. Allday, under cross-examination by Kidwell (4.33A—E) explained that the proposed contract with Japan allowed BNFL to return the spent fuel unreprocessed if attempts at vitrification failed, there being an option to convert the contract into a temporary storage contract. The vitrified wastes would be returned to Japan if vitrification proved successful. (4.32A—C). Allday further explained, under cross-examination, the financial guarantees by the government to BNFL, in the event of the contract not being fulfilled (4.41B—42C).

1.2.23. In response to requests from objectors Silsoe after consultation with Kidwell, read out an agreed summary of the BNFL/Japanese contract, outlining the various 'safety clauses' that protected BNFL.

1.2.24. Allday, examined by Silsoe, considered that one of the major incentives for the development of nuclear power in many countries had been the desire to minimise energy imports, and that the existence of an adequate worldwide supply of uranium was only of relevance if a free market could be assured. He considered that this assumption could not be made due to the growth of 'resource diplomacy'. He argued that for major uranium importing countries the consequent need to minimise uranium imports through the recycling of uranium and plutonium was essential.

1.2.25. Chapman, examined by Thorold, argued that it was unwise to count the economic benefits of overseas earnings from THORP at their full value because their value to the British economy would depend on whether the balance of payments was in surplus or deficit at the time. In other words these earnings would have an adverse effect if the balance of payments was in surplus (e.g. due to exports of North Sea oil) as they might force a revaluation of the £.

1.2.26. Parker (para. 6.25) agrees that 'resource diplomacy' is a potent weapon and commented that to introduce a limitation to reprocessing would prevent the resource independence which is legitimately sought by nations without their own supplies [of uranium].

Case references
Allday 3.44D—45D, 4.32A—C and 33A—E; BNFL 8.16B—19G; Chapman 56.84A—85D

Documents
BNFL 179; FOE 17, 62

1.3. ALTERNATIVE NON-NUCLEAR STRATEGIES: ASSUMPTIONS OF ALTERNATIVES

Issue: Whether or not alternative energy sources could fulfill the energy needs of society after the scaling down or abandonment of a nuclear power programme.

1.3.1 Leach (54. 38A—51D and 55.12C—27B) outlined the alternatives of insulation, fluidised bed combustion, electric heat pumps, and their costs in comparison to nuclear power costs, and argued that all these methods of energy saving could usefully be used in existing housing. He also drew attention to the saving to be made by solar water heating and solar photovoltaic devices. He further argued that as the growth in the number of households slows (as population growth slows and family fission is reduced) past growth trends in domestic fuel consumption will also slow, and that with proper insulation it was perfectly feasible to reduce primary fuel demand for domestic space and water heating to one third of that of comparable new or post World War II housing. (this accounts for 85% of energy consumption in the domestic sector, itself 25% of total national primary fuel consumption). The potential in commercial premises was also seen from various Department of Environment Property Services Agency studies to be considerable, as were the conservation possibilities in industry and transport.

1.3.2. Silsoe in cross-examination (55.28A—84F) argued that energy saving options did not invalidate the need for THORP, and pointed to some of the problems of a massive insulation and conservation programme, such as CHP schemes being no longer economic, and the problems of condensation in some houses. He also argued that of the home insulation programmes outlined in Leach's proof (table 5 p. 20), only the first four steps were marginally cost effective.

1.3.3. Widdicombe in cross-examination of Wright (41.14D—24B) cited BNFL 9 para. 335 on the advantages of an increased investment by the CEGB in CHP. He argued that if 25% of the population could be supplied with CHP about 25 GW less nuclear capacity would be required, a range of 7—30 mtce per annum. Wright countered by citing a Swedish study (no document number cited) which concluded that CHP only saved 1½% of Swedish energy needs. Widdicombe considered the summary of CHP in CEGB 1 to be a very partial statement.

1.3.4. Tweedy (41.57G–H) considered that the Scottish boards could expect a growth rate of 3–4% per annum between 1976/77 and 1986/87.

1.3.5. Taylor cross-examined Jones on the current costs of solar heaters installed in the home. Jones confirmed that no government subsidies were currently given for solar heating installation, and that the government felt more efficient ways of keeping energy costs down could be found.

1.3.6. Page itemised four policies that he considered to be alternatives to an increased nuclear power component in electricity generation, namely, energy conservation, energy storage, expanding coal production, and making effective use of alternative energy sources.

1.3.7. Scargill (35.24G–33B) outlined in more detail the potential role an expanded coal production policy could play, and called for a planned energy programme recognising the part each fuel industry has to play. Under cross-examination Silsoe (66.6C–35B) he reiterated his opposition to all nuclear power, and elaborated his views on the expanded role coal could play in energy supply, calling for an increase in investment and a slowing down in pit closures, arguing that pits should be kept open until all their workable seams had been exhausted. Silsoe questioned Scargill's figures for new investment required, and the rate of new pit openings in order to achieve his projected target of 250 million tons by 1992. The Inspector (66.61B–71A) questioned Scargill at length about the likelihood of achieving a dramatic increase in production and the likely quantity of investment that would be required.

1.3.8. Musgrove argued that wind energy systems could provide at least 25% of our present electricity needs, and that arrangements might best be sited in clusters offshore.

1.3.9. Salter (35.43H–50C) expressed confidence that wave-power could meet a significant fraction of British energy needs by the 1990s.

1.3.10. In cross-examination of Salter (35.92H–108C) Silsoe cited G 3 quoting a figure of 15 mtce as the possible wave-power contribution to energy supplies by 2000. Salter estimated that the maximum possible contribution would be 100 mtce pa and that a figure of 25 mtce could be achieved by 2000.

1.3.11. E. Wilson argued (35.51A–G) that tidal energy could be expected to provide 8% of current electricity consumption if barrages were built on the Solway and the Severn. Silsoe in cross-examination (35.69H–81F) cited BNFL 209, 210 and 211, and argued that an economic Severn barrage was not practicable. Wilson disagreed with this study.

1.3.12. Lord Wilson called for more attention to be paid to the use of hydro-electric power.

1.3.13. Hall for the Society for Environmental Improvement drew attention to the potentials for bio-fuels.

1.3.14. Patterson outlined the concept of fluidised bed combustion, and commended its use in small-scale plants, because of its increased efficiency and environmental advantages.

1.3.15. Chapman (56.60A–81G) outlined the economic evaluation of a range of energy supply systems that have been suggested as replacements for oil and gas in the transport and heating fields.

1.3.16. Silsoe in cross-examination (57.4H–40E) pointed to Chapman's differing predictions on the necessary size of the nuclear component in his paper (BNFL 252) from that in his proof, and suggested that his views as to the required size of the nuclear component might change again, up or down. Chapman agreed but thought the general trend downwards was unlikely to be reversed.

1.3.17. Jenkins (64.1H–11D) outlined the advantages of combined heat and power projects and argued that an energy policy based on CHP would eliminate the need to build more power stations, nuclear or otherwise for at least 20 years. Under cross-examination by Silsoe Jenkins admitted that nuclear power stations could be used in CHP schemes, but suggested that the present British ones were located too far from urban areas to be of use in this respect. Widdicombe, in cross-examination, drew attention to the possibilities of problems of acceptability of hot water from nuclear power stations in CHP schemes.

1.3.18. de Turville (88.81D–87C) outlined the possibilities of utilising organic material in unconsolidated marine sediments in energy production and the methods by which the energy could be extracted as petroleum ether. In supplementary proof (86.7B–8C) he admitted that this scheme, while making a serious contribution to global energy supplies in the next 50–100 years, would in the long term be limited by the buildup of CO_2 in the atmosphere.

1.3.19. Armstrong-Evans (85.87D–90G) advocated the use of an electronic load control system for electricity generation which, he argued, would make more hydro- and wind-electric schemes economic.

1.3.20. Randell (85.91C–96D) outlined the advantages of energy storage in enabling large savings in energy consumption to be made, with particular application to solar heating. Methods advocated included heat pumps, and he also drew attention to the heat-storing properties of some salt hydrates. Ice was also recommended as a principal heat storage medium. Under cross-examination by Tyme (86.2E–3H) Randell said that experiments carried out for Salford Corporation showed that a 50% saving in home heating could be achieved within council house construction yardsticks.

1.3.21. Armstead outlined the various methods of rock shattering in order to utilise geothermal energy. He conceded that its large scale utilisation could not be envisaged before the 21st century, but cited the large scale use of low grade heat in Iceland and Hungary.

1.3.22. Parker comments (para. 8.40) 'As is well recognised by the Government, efforts to effect energy savings by conservation methods, such as insulation or combined heat and power schemes, and to develop alternative sources of energy, such as solar, tidal, wind, wave, biomass or geothermal, should certainly be pursued, but to divert available resources to such efforts to an extent which would prejudice a large-scale reliance on nuclear power should it be needed would . . . be an act of bad management for which this and future generations might justly blame the government . . .'. Later in the same paragraph he said 'Much was made of the ability of coal reserves to deal with any shortfall from other sources in particular by Mr. Arthur Scargill . . . I can only describe his forecasts of what could be achieved as fanciful'.

Case references
Armstead 86.46H–51E; Armstrong-Evans 85.87D–90G; BNFL 35.69F–81H and 92H–108C, 57.4H–40E, 64.11F–18G and 66.6C–35B; Chapman 56.60A–81G, 57.4H–40E; Hall 35.62E–67G; Jenkins 64.1H–11D and 11F–18G; Jones 43.38G–43G; Leach 54.38A–51D, 55.12C–27B and 28A–84F; Musgrove 35.35E–43C; Page 35.12A–24E; Patterson 51.82B–89F; PERG 43.38G–42G; Randell 85.91C–96D and 86.2E–3H; Salter 35.43H–50C and 92H–108C; Scargill 35.24G–33B, 66.6C–35B and 61B–71A; SEI 86.2E–3H; de Turville 85.81D–87C, 86.7B–8C; Tweedy 41.57G–H; WA 41.14D–24B; Wilson 35.51A–G; Wilson, Lord 35.55G–61D; Wright 41.14D–24B; Inspector 66.61B–71A

Proofs
Armstead, Armstrong-Evans, Chapman paras. 2.1–3.5; Leach paras. 71–117; Page; Salter; de Turville proof and supplementary proof; Wilson

Documents
BNFL 9, 120, 176, 209, 210, 211, 235, 236, 252, 253; CEGB 1; FOE 34, 105, 110, 111, 114; G 3; SEI 34; WA 62, 65

ALTERNATIVE NON-NUCLEAR STRATEGIES: RISK ASSESSMENT

Issue: Whether the unconventional* alternative energy sources included in their development or exploitation risks of a different but perhaps as serious a nature as that from nuclear power.

*For convenience we have separated off the 'alternative' of using conventional sources such as coal, upon which there was considerable argument, from unconventional sources such as wind, on which there was little, (cf. 4.1-4.5).

1.3.23. Musgrove argued that the only serious objection to the large-scale use of windmills was their environmental impact especially in areas of outstanding natural beauty, but that their siting in offshore clusters would remove such objections.

1.3.24. Silsoe in cross-examination cited BNFL 215, in referring to the problems of utilising wind power in terms of the considerable impact of aerogenerators and their associated transmission lines upon the visual amenity, particularly in coastal regions.

1.3.25. de Turville in a supplementary proof, admitted that his scheme for the utilisation of organic matter in unconsolidated marine sediments for energy production, while potentially making a serious contribution to global energy supplies in the next 50–100 years would in the longer term be limited by the build up of CO_2 in the atmosphere.

1.3.26. Parker (para. 8.42) citing the Ford Mitre report (BNFL 39) comments that alternative energy sources especially coal carry their own risks, and that in the case of coal these risks include those from radioactive emission (see also 4.1–5).

Case reference
BNFL 36.13B–25H; Musgrove 35.39C–D, 36.13B–25H; de Turville 86.7B–8C

Proofs
Musgrove; de Turville supplementary proof

Documents
BNFL 39, 215; G 3

1.4. SUMMARY OF ASSUMPTIONS AND CASE FOR NUCLEAR STRATEGIES

Issues: (a) Whether or not a nuclear power programme for electricity generation in Britain made economic sense. (b) The role of reprocessing in the nuclear economy.

1.4.1 (a) Jones, in his statement (43.51A–B) said that the general consensus was that an 'energy gap' between indigenous supplies and energy needs would open up before the end of the century. Kidwell, in cross-examination (42.72A–73E) defined an 'energy gap' as being the gap between *indigenous* supplies and needs. Mr. Jones agreed. He further drew his attention to G 11 (42.81A–86E) in which FOE called for the true costs of providing electricity by coal, oil and nuclear fuels, with the accounting basis used for each.

1.4.2. Taylor in cross-examination of Jones (43.32D–37C) questioned whether the Department of Energy felt that the costs of supplying electricity were such that they were having an adverse effect on British economic growth performance. Jones did not think so.

1.4.3. Silsoe drew attention (2.8C–27D) to G 3 outlined the reasons why the Department of Energy favoured the continued development of a nuclear power programme. He further cited BNFL 9 (2.57A–59C) (paras. 189–197) which itemised the advantages of a nuclear programme, both on its own merits, and in reducing environmental and other problems associated with other fuel sources such as coal and oil.

1.4.4. Allday (para. 49) drew attention to the energy and balance of trade savings in the order of 200 million barrels of oil or 50 million tons of coal through reprocessing of spent fuel and the extraction of uranium contained in the 3300 tonnes of irradiated fuel which will arise up to 1995 from the present AGR reactors. He further argued (paras. 56–94) (3.65A–H) that to take on foreign fuel, and to double the plant capacity to 1200 tonnes per annum would only increase total costs by 20% in relation to a plant with a 600 tonne per annum capacity (British arisings only). In apportioning costs of construction etc. on a pro-rata basis this would represent a highly desirable economic strategy. He also itemised the costs of the development at £600 million, and admitted under questioning by the Inspector that this figure did not include interest charges, which enormously increase the figure.

1.4.5. (b) Wright (40.11E–12C) outlined the reasons why the electricity boards thought that the constructions of THORP was desirable, itemising (i) reprocessing being an established method of dealing safely and efficiently with spent fuel; (ii) that it provides a means of recovering valuable uranium and plutonium from the spent oxide fuel; (iii) that it avoids a further commitment to B.205 allowing that to be used for reprocessing Magnox fuel; (iv) that it will provide security for the AGR stations by replacing B.204; (v) that it will help retain the option of a fast breeder; (vi) that it will enable the CEGB to take advantage of reduced reprocessing costs which came from participation by overseas customers; and (vii) that the proposed extent of the CEGB share of the plant would be suitable for either the smallest or the largest nuclear programme that can realistically be envisaged for the rest of the century. He estimated (40.23C–24C) that the costs of reprocessing nuclear waste would vary between 8–17% of total generating costs, depending on whether there was a large or a modest nuclear programme.

1.4.6. Under cross-examination by Kidwell (40.21G–22D) Wright agreed that the 'cost plus' element in the BNFL/CEGB contracts for reprocessing the boards' wastes in THORP were about £250,000 per tonne.

1.4.7. Kidwell further outlined (40.46C–71D) a series of different prices of plutonium using varying discount levels and contrasted this with the non-reprocessing (yellowcake-storage) route. The Inspector agreed that these showed the THORP route to be £350–400 million more expensive than the storage route, but pointed out that this was only part of the calculation, having not gone anything like the full route. Kidwell replied that in order for the storage route to become as expensive as the reprocessing route, uranium prices would need to rise to $100 per lb. Wright feared that storage might cause the CEGB to store their fuel in their own storage ponds while further storage capacity was constructed, and as their storage space was limited it might entail shutting down the stations at a cost of £300 million per annum, and at the risk of serious power cuts.

1.4.8. Avery outlined the costs (56.2E–16G) in BNFL 232 and 232A, of dealing with 3000 tons of AGR irradiated fuel, arising from the CEGB/SSEB reactors, over the period 1977/78 to 1994/95 in respect of four alternative strategies of disposal. Kidwell in cross-examination (56.17A–35G and 50C–53D) questioned many of the figures and the ways at which they were arrived at for the non-reprocessing (yellowcake-storage) route, especially the methods by which costs were discounted.

1.4.9. Chapman argued that for a number of reasons outlined in his proof 'an economic case for the reprocessing plant does not yet exist'.

1.4.10. Sweet criticised the CEGB's methods of calculating generating costs in their nuclear stations, specifically their discounting procedures, (see BNFL 232 for discounted costs) arguing that in the long-term nuclear power stations would cause a significant increase in the price of electricity, and that factors such as research and development, and decommissionning costs are not taken into account when calculating the unit costs of nuclear generated electricity. He further criticised their reluctance to release figures to enable people outside the industry to carry out their own calculations of the CEGB's generating costs. He also criticised BNFL for failing to make a proper economic assessment of the proposed THORP project.

1.4.11. Parker comments (para. 9.5) 'from the evidence produced it is clear that viewed simply as a plant for the production of uranium and plutonium, the plant would not break even unless and until the price of uranium increased very markedly against reprocessing costs . . . viewed purely as a production plant it appears to me that THORP would be economically disadvantageous in absolute terms'. He does say (para. 9. iii) that 'the financial disadvantage might be an acceptable price for some other advantage, for example, resource independence, reduction of plutonium stocks, or anti-proliferation effect'.

Case references
Avery 56.2E–16G, 17A–35G and 50C–53D; BNFL 2.8C–27D and 57A–59C; Chapman 56.88F–90F; FOE 31.78G–95A, 40.11E–12C, 21G–22D and 46C–71D, 42.72A–73E and 81A–86E, 56.17A–35G and 50C–53D; PERG 43.32D–37C; Sweet 65.1E–38E; Wright 40.11E–12C, 21G–22D, 23C–24C and

Proofs
Allday paras. 49–59; Chapman para 96; Sweet paras. 1.3–3.20; Wright

Statement
Jones

Documents
BNFL 9, 170, 232, 232A, 265; FOE 101, 102, 103; G 3, 11

2. Weapons Proliferation

2.1. INTERNATIONAL AGREEMENTS: NON-PROLIFERATION TREATY

Issue: whether or not the proposed use of THORP to reprocess spent foreign fuel broke either the spirit or the letter of the Non-Proliferation Treaty.

2.1.1.   Silsoe drew attention to the fact that the government was satisfied that nothing in BNFL's proposed contract with Japan would be in breach of the UK's obligations under the NPT. Allday assured the Inspector that the companies financial position would be protected though the business and profit would be lost if there was a US imposed ban on the transport and reprocessing of fuel of US origin at the proposed plant. Avery outlined the provision of the NPT and Euratom treaties. Kidwell drew attention to FOE documents 53, 88 and 89 on the Canadian attitude towards reprocessing and the sale of uranium to non-NPT countries. He quoted Ranger I for the Australian position, the Flowers Report and the May 1977 White Paper (BNFL 170) for the British.

2.1.2.   He further commented in his closing speech that the distribution of plutonium in any form to non-weapons states was undesirable and that all the legal interpretations of the Non Proliferation Treaty could not force Britain to build something (THORP) which would be directly contrary to the spirit of the NPT by putting the raw material for nuclear weapons in the hands of a non-nuclear weapons state.

2.1.3.   Oakes outlined the extent and limitations of the NPT and the various bilateral agreements between the USA and the USSR, and argued that in some cases these agreements raised rather than lowered the levels of nuclear armament.

2.1.4.   Cochran outlined the limitations of the current international agreements.

2.1.5.   Wohlstetter criticised the ambiguities of the NPT and felt strongly that Article IV did not mean that non-nuclear powers were entitled to stocks of plutonium, in exchange for promises that these would not be used to make a bomb.

2.1.6. Johnston criticised the ineffectiveness of the NPT and the Euratom Treaty, and claimed that their shortcomings were sometimes used as an excuse for nuclear exports.

2.1.7. Rotblat criticised the NPT for implicitly condoning research and development leading to the manufacture of nuclear weapons, and criticised the IAEA system for having no authority to exercise any policy-type activities, or to provide physical security or prevent theft or sabotage. He called for a strengthening of the safeguards and argued that the greatest peril would be a 'plutonium economy' in which the world depended on FBR's as a major source of energy.

2.1.8. Parker (para. 6–15 (2)) comments 'I find it difficult to see how a party which has developed reprocessing technology or created reprocessing facilities, could be otherwise than in breach of the agreement (NPT), if it both refused to supply the technology to another party, and refused to reprocess for it'. He continues (para. 6–33) 'The argument that the grant of permission would add to proliferation risks was not however established before me. Indeed I would go further. Since (i) there will be no direct risk arising from THORP for at least ten years (ii) to deny reprocessing facilities would be against the spirit—and as I think the letter—of our obligations under the main existing bulkward against proliferation (NPT) . . . I do not accept that the best way to achieve a new bargain is to break an existing one'.

Case references
Allday 3.67H–68C; Avery 30.40A–42C; BNFL 2.42C–43B; FOE 30.93A–106F and 93.33C–D; Oakes 61.22D–28D; Rotblat 74.86D–89F; Wohlstetter 58.38D–46A.

Proofs
Allday Addendum 1 to proof; Avery paras. 2–11; Cochran paras. 5–13; Johnston paras. 1–20; Oakes paras. 8.1–12.7; Rotblat paras. 58–78; Wohlstetter paras. 44–46. (Note: the relevant sections of the proofs of Cochran and Johnston were not read into the transcript.)

Documents
BNFL 31, 49, 51, 170; FOE 29, 53, 88, 89, 116; NPC 39; TCPA 9

INTERNATIONAL AGREEMENTS: PRESENT AND POTENTIAL NUCLEAR POWERS

Issue: Whether or not the construction of THORP would assist in the spread of nuclear weapons potential to non-weapons states.

2.1.9. Silsoe drew attention to the Flowers Report's (BNFL 9) view on weapons proliferation and expressed the view that countries determined to develop nuclear weapons could do so whether reprocessing was banned or not. He also drew attention to President Carter's 7 April 1977 statement and subsequent questions on it, and argued that the best means of discouraging proliferation as President Carter had requested lay in strengthening the safeguard system and not by abandoning reprocessing at Windscale.

2.1.10. Kidwell in cross-examination of Herzig (44.8B–19H) discussed the possibility of American intervention in the THORP contract stopping the Japanese from sending their spent fuel, and argued that the financial position of THORP was therefore uncertain. Herzig agreed that there was doubt about what American policy would be in 1982 but insisted that BNFL were adequately financially protected.

2.1.11. Alesbury in cross-examination of Herzig (44.9D–15E) drew attention to the Euratom Treaty (BNFL 50) and argued that confidence in nuclear matters was now less than when the treaty was first drawn up. He argued that the existence of Euratom as an institution created a pro-nuclear bias in the European Commission. Herzig countered by arguing that the oil crisis had had a more profoundly pro-nuclear effect on the commission, and cited the existence of the coal and steel community as evidence of an equally pro-coal bias, on the same analogy.

2.1.12. Kidwell (30.60A–106G, 31.1B–30C) in cross-examination of Avery defined non-proliferation as being concerned with preventing those who do not at present have a nuclear weapons capacity from acquiring one. He drew attention to FOE 37 and questioned Avery about his views on the possibilities of making nuclear weapons from reactor grade fuel. Avery agreed that it was possible. He further drew attention to FOE 77, 82 (Nye Statement) and BNFL 32 (Carter's statement). Avery was concerned that the impression given by these statements was of American myopia in assuming that the current energy mix in the US implied that all other countries were in the same situation. The inspector pointed out to Kidwell that President Carter's statement included the caveat that those who already had facilities for reprocessing could continue with it. Kidwell further cited FOE 87 and 90 which refer to the US policy concerning the re-transfer of US-origin spent fuel for reprocessing on a case by case basis.

2.1.13. Oakes criticised the present nuclear powers for their lax attitude towards nuclear proliferation, and pointed to what she saw as a direct relation between nuclear weapons and nuclear power production.

2.1.14 Wohlstetter (58.36E–37B) pointed out that contrary to industry claims, it was possible to make explosive devices with reactor grade plutonium, and that the US government had exploded one. He further (58.46C–47G and 59.3B–12G) attacked the economic mismanagement by nuclear powers of non-nuclear powers 'peaceful' nuclear projects, and suggested that more vigorous costings would have shown that such peaceful projects were only economic if they had military spinoffs. He cited the cases of Israel (whose research reactor was called a textile plant), India (who violated their agreement with the US by using the heavy water supplied to extract plutonium) and Pakistan (who went to great lengths to insist on the fulfilment of its contract with France for a nominally civilian reprocessing facility, while at the same time offering to cancel its deal with France if the other nuclear powers guaranteed to destroy their weapons), and argued that deliberate ambiguities enabled bureaucracies to inch towards a nuclear capability often without the rapidly changing political leaders realising. Wohlstetter also argued that such ambiguity between civilian and military application also assisted in keeping the decision covert once decided upon, citing the Indian case.

2.1.15. Parker (para. 6.33) comments that 'The argument that the grant of permission would add to proliferation risks was not however established before me'.

Case references
Avery 30.60A−106G and 31.1B−30C; BNFL 2.62H−72D; FOE 30.60A−106G, 31.1B−30C and 44.8B−19H; Herzig 44.8B−19H; Oakes 61.16A−22C; WA 44.9D−15E; Wohlstetter 56.36E−37B, 46C−47G and 59.3B−12G

Proofs
Oakes paras. 1.1−7.9; Wohlstetter paras. 37−39 and 47−86

Documents
BNFL 9, 32, 50, 53; FOE 28, 37, 77, 86, 87, 90; NPC 38, 41, 43, 45, 49, 57

2.2. PLUTONIUM: RISKS AND SAFEGUARDS

Issue: Whether or not the return of plutonium, after reprocessing to non-nuclear weapons states, would increase the chances of weapons proliferation, or whether it was possible to avoid this (e.g. through 'spiking' of the returned material).

2.2.1. Rotblat drew attention to the risks of plutonium being acquired by governments to manufacture nuclear weapons, and argued that the more obstacles placed in the way of this the less likely was it to happen.

2.2.2. Wohlstetter in examination-in-chief by Kidwell (59.27A−30E) argued that if THORP were to go ahead it would legitimise commerce in separated plutonium and make it extremely difficult not to sell separated plutonium to Korea, Taiwan, and other unstable and/or undemocratic governments. He argued that a delay of ten years, on the other hand, would be a major influence on other nations such as France and Japan in deciding their nuclear policies. Under cross-examination by Silsoe (59.62D–72A) he outlined the arguments in FOE 28, pointed to the small difference between stages two and three (small bomb manufacture, which could be used to threaten non-nuclear neighbours, and larger-scale military deployment of fifty or so bombs, possible only a couple of years later), and argued that multinational centres for the making and distribution of fresh mixed plutonium and oxide fuels would not help solve the proliferation problem.

2.2.3. Widdicombe in cross-examination of Allday drew attention to Gilinsky's paper (FOE 37) on the dangers of the illicit diversion of power producing plutonium into nuclear weapons materials by governments. Gilinsky refers to President Ford's 1976 statement, and Allday considered that this statement would be unlikely to help non-proliferation, though he supported that part of the statement which called for a more rigorous control of reprocessing technology.

2.2.4. Kidwell drew to the attention of Avery those paragraphs in the Japanese contract (BNFL 179) referring to plutonium (para. 11); an exchange between Silsoe, Kidwell, and Widdicombe revealed that there was some uncertainty as to whether this contract allowed the 'spiking' of fuel before its return to Japan.

2.2.5. Kidwell further commented that he did not accept that transport in the form of oxide would be any safeguard against material being taken and used. He argued that to irradiate fuel was expensive and that costs would be further increased by the need to take account of the design of each reactor for which material would be irradiated, and further agreed that as regards the Japanese contract, such an idea would not be possible in any case.

2.2.6. Parker (para. 6.30b) commented that extra reprocessing facilities would need to be created on fuel management grounds and their creation in present nuclear weapons states would not increase the proliferation risk unless the plutonium produced by the facilities were returned to fuel owners in a form which would enable the owner country to proceed to a bomb without time for diplomatic pressure to be exerted. He agreed (para. 6.32) that returning the plutonium to non-nuclear weapon owning countries would represent an increased risk, but that this might be mitigated by returning only when required for civil reactors, and then only in the form of briefly irradiated fuel rods.

Case references
Allday 6.91H–94E and 7.7A–10C; Avery 30.64C–67D; BNFL 30.67D–68G

and 59.62D—72A; FOE 30.67D—68G and 93.6E—7B; WA 30.67D—68G; Wohlstetter 59.27A—30E and 62D—72A

Proof
Rotblat paras. 54—57 (this part of his proof was not read into the transcript)

Documents
BNFL 39, 51, 179, 196; FOE 28, 37, 118; TCPA 23

PLUTONIUM: PROVISION OF REPROCESSING FACILITIES

Issue: Whether the offering of reprocessing facilities to other countries increased the chances of weapons proliferation, and whether international safeguards ensured that the plutonium was not diverted to weapons use.

2.2.7. Silsoe (2.50G—56D) drew attention to the international safeguards against plutonium diversion by terrorists as outlined in the Flowers Report, and to the fears expressed in that report of the risks of proliferation in having larger quantities of plutonium in world circulation as a result of reprocessing.

2.2.8. Under cross-examination by Kidwell, Allday expressed the view that the increased availability of plutonium as a consequence of reprocessing did not not significantly increase the likelihood or possibility of another nation obtaining weapons potential. Allday saw the problem of proliferation as being one of the uncontrolled dissemination of technology, rather than of plutonium.

2.2.9. Kidwell countered this view (8.9B—17D) by pointing to the Flowers Report (para. 532, sub para. 23) and the Ranger Report I, which recommends that plutonium should not be made available to countries that are not party to the NPT. He also presented FOE 11, 28, and 37 as further documentation of this position.

2.2.10. He also drew attention to Avery's proof (31.9C—26E) on the subject of safeguards, materials unaccounted for (MUF), containment and surveillance, being itemised as the main IAEA safeguards, MUF being the primary element. He further drew attention during cross-examination of Avery to BNFL 49 and FOE 12 (which is criticisms of BNFL 49) and Avery agreed that IAEA inspection was tokenal as the agency was grossly understaffed. (Avery in his proof para. 8 says 'The purpose of safeguards systems can only be to make diversion of material from the civil nuclear fuel cycle to weapons uses, as difficult to conceal and as politically damaging as possible; they cannot prevent a determined nation from independently acquiring a nuclear weapons capability:) Johnston drew attention to the technical and political shortcomings in the international nuclear safeguards machinery, particularly the IAEA safeguards. He considered the Euratom inspection system to be superior, but nevertheless inadequate, in not including surveillance or escort of shipments of uranium or other fissionable materials (for example it failed to prevent the diversion to Israel of 200 tons of uranium material in 1968)

2.2.11. Wohlstetter outlined his reasons for considering that the development of an FBR programme would not require the construction of THORP at this time, citing FOE 120 as substantiation of this submission that the commercial operation of FBRs would not be until about 2000. In cross-examination by Silsoe (59.31A–47G and 49B–61A) he was questioned on the right of countries other than the US to engage in reprocessing and the development of FBRs.

2.2.12. Rotblat under cross-examination by Silsoe outlined a long term policy for reducing the chances of nuclear was through the reprocessing of nuclear fuels, including (1) the tackling of nuclear energy as an international enterprise, rather than as a commercial operation because of its links with nuclear weapons; (2) the establishment of an international bank for the aquisition, storage and allocation of nuclear fuels; (3) the abandonment of reprocessing in favour of storage; (4) research and development into the FBR to be carried out on an international basis in a single project, as an energy 'longstop'; (5) a massive programme of research into alternative energy sources financed and co-ordinated by a UN agency.

2.2.13. Parker (para. 6.6) agrees that the system of safeguards is basically one of reporting and inspection, and remarks that, 'It is sufficient to say that it could and should be improved . . .'.

Case reference
Allday 6.6B–7G; Avery 31.9C–26E; BNFL 2.50G–56D, 59.31A–47G and 49B–61A; FOE 6.9B–17B and 31.9C–26E; Rotblat 75.8E–10E; Wohlstetter 58.14F–17B.

Proofs
Avery paras. 3–8 and 20–21; Johnston paras. 1–20; Wohlstetter para. 21 (Note: Johnston's proof was accepted as a document; it was neither read into the transcript, nor cross-examined.)

Documents
BNFL 9, 10, 39, 49, 262; FOE 11, 12, 15, 16, 28, 29, 37, 45, 90, 116, 119, 120; TCPA 96; WA 50

2.3. ALTERNATIVE FUEL CYCLES

Issue: Whether alternative methods of disposal of spent nuclear fuel, apart from reprocessing, were more or less likely to lead to weapons proliferation.

2.3.1 Cochran NRDC, argued that co-processing, or co-precipitation were not the answer to the proliferation risks of reprocessing (cf. 3.3.4).

2.3.2. Wohlstetter criticised BNFL for not considering the direct disposal of wastes, rather than reprocessing, and cited Canadian American and Swedish experts who considered that disposal of unreprocessed spent fuel in stable geological formations was perfectly safe. He argued that the results of these studies favoured a decision to defer the expansion at Windscale.

2.3.3. Parker, in the section of his report devoted to the nuclear weapons proliferation question (paras. 6.1–6.34) makes no specific reference to alternative methods of disposal and their relative likelihood of increasing or reducing the chances of nuclear weapons proliferation.

Case reference
Cochran 63.9A–10B; Wohlstetter 58.20E–32E

Proofs
Cochran paras. 23–39; Wohlstetter paras. 26–33

Documents
FOE 43, 68, 84, 85; NRDC 2, 3

3. Security and Civil Liberties

3.1 SECURITY AND THE CONTROL OF TECHNOLOGY POLICY

Issue: Whether the degree of secrecy involved in the assessment of hazard potential at Windscale prejudiced the public acceptability of the project.

Accident Analysis

3.1.1. It was argued by Thompson, (cf. 4.3.79–98, 4.5.115–122), that the criterion 'acceptably safe' by which Windscale was to be judged (evidence of the NII, and see Parker, chapter 11), could not be met without thorough and open safety studies. Thompson's proof gave an outline of the extensive nature of such studies. In the course of the Inquiry BNFL made it clear that they would not undertake the kind of analysis proposed by Thompson because it was not in their view a credible hazard potential. When asked to provide sufficient data for PERG to carry out an alternative hazard analysis, Parker ruled that information should be provided 'subject to security considerations'. In the event, no detailed technical data on HAW or spent fuel storage was provided, (89.106B). In addition no analysis of accident mechanisms was possiboe for the plutonium finishing line (Smith 16.85, Donoghue 24.19–20, 25.1–2).

3.1.2. As detailed in paras. 4.3.79–98 below PERG argued that Windscale had a hazard potential far in excess of that admitted by BNFL. Detailed examination of this hazard potential was not possible for security reasons, in particular that engineering detail which could be of use to groups of evil intent, would not be released. PERG argued that this prevented comparison of nuclear energy options with conventional or alternative options from the standpoint of safety and risk to the public and hence made the nuclear option unacceptable by the criteria put forward by the authorities concerned to assess safety.

Discharge Control

3.1.3. A major issue concerning the control of discharges of low-level liquid waste to the Irish Sea arose with regard to the technology for retaining alpha-emitters such as plutonium and americium. NNC and PERG and WIERC attempted to identify the major source of the pollution and, as detailed in 4.3.42–46, it became apparent that the major part came from the weapons programme in plant adjacent to the civil reprocessing complex. In addition data on discharges prior to 1957 and to some extent prior to 1971 was subject to security clearance. Hookway for DOE, argued that security cover was necessary because, 'one could reveal trends' and from a security point of view 'trends are as important as absolute facts' (56.61GH, 62–63, 66). With regard to Plutonium discharges, Taylor questioned Avery but was told by the Inspector that where this matter was concerned, there was an unavoidable dilemma, 'how can we be sure unless we are told, how can we be safe if we are told' (31.86–87, see also Warner 30.14D).

3.1.4.   BNFL countered that discharge control operated at the outlet for the whole site and that therefore the Defence establishment discharges were subject to the same overall control as the civil programme.

Case references
Avery 31.86–87; Donoghue 24.19–20 and 25.1–2; Hookway 56.61G–H, 62–63 and 66; PERG 30.14D; Thompson 89.106B; Warner 30.14D.

Proof
Thompson

3.2. SECURITY AND THE RIGHTS OF THE WORKFORCE

Issue: Whether the technology of THORP and related Windscale facilities would lead to an erosion of workers' rights, to a worsening of industrial relations and to greater safety hazards.

Safety and Strike Action

3.2.1.   It was contended, principally by SERA, that expansion of the Windscale site would add to a progressive criminalisation of what was elsewhere regarded as legitimate industrial action (Lewis 82.26B–D). It was argued that the necessity of troops to safeguard supplies essential for maintenance of safety, presented a potential hazard in that disputes could excalate because in practice agreements on essential supplies were difficult to achieve. The problems of secrecy and information to the workforce exacerbated the situation. SERA noted that the TUC had supported the expansion of Windscale and had also maintained that there was no threat to workers' rights inherent in the expansion, they argued, however, that there was an 'inconsistency' in this approach (82.1E and 3C, TUC 1, 2).

3.2.2. Lewis, an industrial relations expert, argued that the potential for worsening industrial relations and its implications for safety was great. Adams for the EETPU countered this view, presenting evidence of the facilities enjoyed by the Unions at Windscale. Lewis questioned the representativeness of the 'official' union view in the light of the unofficial strike early in 1977 with its rumours of troops and strike breaking. Adams assured Lewis that they had said emphatically to the Secretary of State during his intervention in the strike, that 'there was no question of the use of troops' (82.27D–E). Elsewhere he admitted that the strike was unofficial and that the 'safety weapon' was used and 'that they took the management to the brink is without doubt' (66.106G). He also admitted that there had been dispute as to what constituted essential supplies for plant safety (66.93).

3.2.3. Robertson, a former engineer at Windscale and an individual objector, argued that where such extreme hazards to public safety existed, as at a reprocessing plant, the right to strike by the workforce should be restricted and that only the police or armed forces had the necessary discipline to guarantee public safety (48.31–36). Shortis argued that THORP was now only at the design stage and that his personal opinion as designer, was that there would be no need to restrict the right to strike, (16.43G). (It was not clear if this also applied to related facilities such as HAW tanks.)

Parker's conclusions are contained in paras, 11.19–11.23 of his report. His conclusions can best be listed. (i) He accepted Robertson's experience as relevant (para. 11.20). (ii) He does not report Adams's statement that the workforce took the management to the brink, but states that newspapers had suggested the fact (para. 11.20). (iii) He reports BNFL's, and the TUC/Adams view that restriction on the right to strike is unnecessary and that there would have to be agreed procedures to guarantee safety. (iv) He states that 'agreed procedures' are impossible to guarantee. (v) He argues that if the absence of the workforce would be likely to create significant hazards then he would endorse Robertson's view, and adds that he does not consider this likely provided essential supplies can be guaranteed. This guarantee can only be assured by the use of troops (para 11.22). He acknowledges that this could escalate dispute and that there is a potential for misjudgement of the time when troops 'could be held off no longer'.

Case references
Adams 66.93–106G; Lewis 82.26B–D and 27; Robertson 48.31–36; SERA 82.1E and 3C; Shortis 16.43G.

Proof
Lewis

Documents
TUC 1, 2

3.3. SECURITY AND THE PROBLEM OF TERRORISM OR SABOTAGE

Issue: Whether the secrecy that necessarily surrounded security measures against persons of evil intent led to the impossibility of public assurance on safety.

3.3.1. Questions were raised by several parties on the adequacy of security measures with regard to nuclear material such as plutonium in transit, spent fuel in transit, plutonium and uranium in store, and with regard to potentially hazardous installations such as the HAW tanks and spent-fuel storage ponds. In general, because of security implications these questions could not be answered, nor assertions cross-examined. We therefore reference here some of the avenues followed, albeit but shortly.

3.3.2. Pedlar for WA provided a proof which maintained that Windscale was not secure against armed attack (84.94). The proof was not read into the transcript nor cross-examined upon at the request of the Inspector. Smith for BNFL gave evidence on plutonium storage but was not able to disclose the nature of the plutonium monitoring system by which thefts from the store could be prevented (16.93). Kidwell referred to the difficulties of inspection of such safeguards as noted by the RCEP (16.98, BNFL 9 para 329). Widdicombe referred to the US incidents surrounding the 'Silkwood' case and the UK experiences in the Fuchs and Nunn-May case (85.77–78, WA 185 p. 182, WA 184). Widdicombe also requested the Board of Inquiry Report on the 1970 criticality incident, but was advised against such a request by the Inspector (16.94E). On the question of Plutonium traffic Alesbury was prevented from detailed questioning (24.49) and remarked, 'what is secret is not necessarily safe' (27.50).

3.3.3. On the more general question of terrorism, it was accepted by all sides that such groups would be able to make a credible nuclear weapon. It was also accepted that this could be made from 'reactor-grade' plutonium. At dispute was the ability of terrorists to get hold of fissile material. No evidence could be presented on the security of the site. With regard to material in transit, no evidence could be submitted on plutonium movements, except to state that plutonium had been sold to a number of countries (Italy, Japan, members of Euratom, in particular West Germany), and that so far there had been no problems (Milne 27.45). However, with regard to fuel movements, it was stated that no special precautions were taken. It was considered that the weight of the container, and in the case of spent-fuel, the radiation hazard, were effective barriers against hijack. It was acknowledged by all sides that in order to obtain plutonium from spent-fuel, terrorists would require access to laboratory reprocessing facilities as well as fuel-handling machines and that in this case the prime risk arose from States carrying out diversion of material (cf. 2.1.10–2.1.18).

3.3.4. Thus with regard to diversion of materials for weapons, spent nuclear fuel was not considered a great hazard, and stored more 'accessible' material such as plutonium, was subject to a security blanket. The case of Mixed Oxide Fuel, or MOX, was raised. This contains plutonium but as a fresh fuel is not radioactive. It would thus have a higher potential for diversion en route, particularly as it would be supplied to foreign countries and be an integral part of a FBR reactor programme. Cochran for NRDC was questioned by Parker on the possibility of 'spiking' or irradiating the fuel, such that it presented a radiation field to would-be highjackers. No evidence was presented on this, however, Cochran felt that this would eliminate the terrorist threat but not the State diversion hazard (63.28F). He also stated that it would be difficult to engineer and to gain agreement between customer and supplier and the matter was currently subject to international evaluation. Parker concluded that it could eliminate the terrorist threat, but would be costly and difficult to impose (para. 7.7).

3.3.5. Several parties argued that the public could not be assured on matters of security and therefore that THORP could not be viewed as an 'acceptable' technology. In addition, questions were raised with regard to the security of current activities at Windscale. Justice, the British section of the International Commission of Jurists, called for an independent Security Commission to vett arrangements at Windscale, and WA called for such a special inquiry before THORP could be given planning permission (92.48).

3.3.6. Parker recognised the problem, but felt that powers for checking and reviewing security procedures already existed (para. 7.17–18). He agreed that a check and regular review should take place, but that there was no solution to the problem of assurances (para 7.22–24). He also noted that as THORP was not at the complete design stage, and that design could account for a great deal of security, such an evaluation could not take place at the 'outline' stage now before him.

Case references
Cochran 63.28F; Justice 92.48; Milne 27.43; Pedlar 84.94; Smith 16.93, 94E and 98; WA 24.49–50 and 85.77–78

Proofs
Milne; Pedlar; Smith

3.4. CIVIL LIBERTIES AND NUCLEAR EXPANSION

Issue: Whether THORP and the expansion of nuclear power would lead to an erosion of civil liberties.

3.4.1. Argument on this issue can be separated into that referring specifically to THORP, and that referring to a future 'Plutonium Economy'. The second was necessarily more diffuse, and of questionable relevance according to Parker, who did not accept that THORP involved any such commitment or consequence (see Chapter 7). With regard to THORP itself, we have already referenced the argument concerning the right to strike in section 3.2. In addition, reference was made to current security measures as already infringing the rights of individuals and we deal with this briefly below.

THORP

3.4.2. Several parties advanced the view that current security measures such as positive and negative vetting, restriction of information etc were examples of a progressive erosion of civil liberties concurrent with nuclear developments (Lewis 82.19—20, Grove-White 83.42, FOE 11, Sieghart 83.16—20). Adams for EETPU and the TUC maintained that there was no present infringement of rights as far as the workforce was concerned. Bartlett, for BNFL quoted the RCEP (BNFL 9, paras. 308, 335), maintaining that at present the risk of infringement was small. Sieghart added that perception of what was an infringement would depend on the problem of unemployment in a particular area (83.33D).

3.4.3. An additional area of civil rights related to the THORP development was outlined by Kidwell, Glidewell and Sieghart (proof para. 7—15, 83.4—5), following questions by Taylor with regard to the Nuclear Installations Act, (PERG 25 section 7, PERG 17 question 2, PERG 18 question 2b). Taylor noted that the act placed an absolute liability on the licensee to secure that, '(a) no occurrence . . . causes injury to any person, (b) or any waste discharge causes injury to any person'. Parker ruled that questions seeking to define 'injury' were out-of-order as they were points of law. Glidewell offered to clarify the situation, but provided only a precis of the Act, and Kidwell remarked that it would be impossible for a claimant under the act to prove liability as the injury would be a cancer which developed many years after and with only a certain probability of having been caused by the 'occurrence' or the discharge. Sieghart's summary of the situation was countered by Bartlett for BNFL who argued that it was not proven that other energy sources did not constitute similar threats and it might well be the case that nuclear energy supply options reduced the overall cancer hazard (83.29).

3.4.4. Parker concluded that the Act did indeed offer only 'illusory' protection, but that in the light of other risks, for example from smoking, he could not recommend a change in the law (paras. 10.123−124).

The Plutonium Economy

3.4.5. Several parties advanced the view that THORP should be viewed as a first step on the road to a 'Plutonium Economy', i.e. a fuel economy in which plutonium has become a major item of commerce (WA, NCCL, Justice, FOE, PERG). It was countered by BNFL that plutonium was already an item of commerce, there having been shipments of plutonium, and Pu-containing fuels, and that a certain amount of MOX had been used in thermal reactors. BNFL also argued that THORP although allowing the FBR option to remain open, was not necessary for the early stages and did not pre-determine an FBR programme. Parker concluded that THORP did not presuppose a large scale nuclear expansion (paras. 7.1−17).

3.4.6. The point is important because parties gave great weight to the views of the RCEP, which concluded that the dangers due to sabotage, detonation of a nuclear device, dispersal of toxic nuclear material or political and economic demands from such blackmail, and the threat to civil liberties arising from measures taken to contain the danger, were the most worrying aspects of nuclear expansion (BNFL 9, para 505). Sieghart emphasised this view, referring to the 'charisma' of Plutonium, and concluding that 'no other energy source presents comparable security problems' (83.26D).

3.4.7. Little evidence was presented on the potential loss of civil liberties. A question was asked by PERG on surveillance of opponents to the nuclear programme (PERG 18 question 1). This was answered by the Department of Energy, stating that opponents of nuclear power would not be put under surveillance unless they were considered 'subversive, violent, or otherwise unlawful'. Herzig for DEn was asked to define the term 'subversive' and referred to a definition given in a House of Lords debate: 'activities threatening the safety or well-being of the State and intended to undermine or overthrow Parliamentary Democracy by political, industrial or violent means' (44.16).

3.4.8. Sieghart stated that Justice was most perturbed at the wide interpretation of the term. It meant that surveillance could be countenanced where industrial action sought to affect parliamentary decision-making, and the breadth of 'political' means could readily include all forms of pressure groups, (83.15−16). In addition it was noted that the UKAEA had its own armed police force with powers of search and arrest beyond the confines of particular installations (83.10−12).

3.4.9. This argument was also put by NCCL, who regarded THORP as a first step to the Plutonium Economy (95.20G−H). Their witness was particularly concerned with the erosion of civil liberties caused by the need for vetting and surveillance when Plutonium moved from the military to the civil sphere (Grove-White 83.42). Blom-Cooper, argued in closing speech that 'files will be opened on opponents to nuclear development. They will be subject to telephone taps, mail interference, opening of bank accounts and other invasions of their privacy' (95.25F−H), and that the possibility of a 'holocaust' would always tip the balance against human rights. He referred to the Hosenball case where a judge had ruled that 'when the State itself is in danger our cherished freedoms have to take second place. Even natural justice itself may suffer a set back' (95.26).

3.4.10. Parker did not accept that THORP necessarily meant a commitment to a FBR programme, or to an expanded reactor programme of the kind that had worried the RCEP. However, he did accept that there was a problem relating to the protection of human rights where such an expansion was concerned. He concluded that there was no solution to the problem if the country was to be protected (paras. 7.22−24).

Case references
Grove-White 83.42; Herzig 44.16; Lewis 82.19−20; NCCL 95.4−5, 20−26 and 29; Sieghart 83.10−12, 15−20 and 33D

Proofs
Grove-White; Sieghart

Documents
BNFL 9; FOE 11; PERG 17, 18, 25

4. Health and Safety

4.1. COMPARATIVE ASPECTS OF THE REPROCESSING OPTION

4.1.1. The evidence and argument on matters of health and safety ranged widely. No one party presented a detailed critique covering all the main themes and many points were raised which were not later tied in to a specific position. Parker drew out three main themes to the arguments: (i) what was to be done with the waste, i.e. should it be reprocessed as part of waste management and/or fuel conservation; (ii) what were the relative health and safety consequences of the options; and (iii) could BNFL and the authorities be relied upon to meet intentions (paras. 2.20–29, 10.18).

4.1.2. However, there is a problem in following this schema, in that very little evidence was tendered on comparative aspects. Parker, for example, refers to the health penalties of non-reprocessing, arguing that there would be increased uranium mining as a result, with concomitant health impact. Likewise, he states that nuclear generating stations are generally recognised as 'cleaner' than conventional coal stations in terms of the health impact of emissions. He quotes the Ford Mitre Report in support of this contention (para. 8.40–42 and BNFL 39). Yet at no point were any detailed figures produced comparing alternative fuel cycles. Taylor held that no conclusions could be drawn on the basis of the evidence submitted to the Inquiry (96.16–17) and Spearing had pointed out that the Ford Mitre Study had concluded that reprocessing exacerbated problems of waste management, and in addition that its conclusion with regard to coal stations versus nuclear was based on the assumption that the fuel would not be reprocessed (when the bulk of the emissions-per-gigawatt year take place). He further added that recent commentators had pointed to the health hazard from radon gas released from uranium tailings as being a significant long term source of health effects for comparison with coal (88.4–6, 7, 97.15–17).

4.1.3. In spite of the evident lack of data on this question, no parties called for an adjournment on the matter, and only PERG presented recommendations for future health and safety research, holding that the application should be refused until such time as effective comparisons could be made (PERG 80, 81).

4.1.4. The bulk of the criticisms regarding health and safety related rather to the likelihood of the applicant meeting the stated intentions with regard to the impact on the workforce, the local community and wider afield. This critique took the following avenues. (i) The past history of the Magnox plant discharges and control technology was scrutinised. (ii) The present position with regard to the results of those discharges, in particular the build-up in the local environment of concentrations of particular nuclides, was investigated in some detail. (iii) The standard upon which the significance of past and present activities were assessed was questioned, particularly with regard to the assessment of risk from low doses of radiation received chronically. (iv) The knowledge upon which present models of the behaviour of radionuclides in the environment was based was questioned, as in a number of cases there had been no direct measurements to validate calculations.

4.1.5. Thus the proponents tended to base their case on an incremental logic—that reprocessing was an established technology and the only proven means of waste management; that it was rational energy conservation; and that there were no objections from the authorising bodies on matters of health or safety. Parker sought to introduce the comparative component. Whereas the objectors were primarily concerned with (a) a re-assessment of the present health impact; and (b) the likelihood of the applicant meeting the intended environmental impact targets.

Summary of Arguments on Health and Safety

4.1.6. BNFL maintained that reprocessing was the only established method of control of radioactive wastes. The processes had been carefully vetted by the authorities, the Nuclear Installations Inspectorate, and the discharges carefully monitored by the Ministry of Agriculture Fisheries and Food. The standards by which the authorities judged the acceptability of the discharges were reviewed by an international body and at a national level, the system having been approved by the Royal Commission on Nuclear Power and the Environment (1.8, 2.18, 2.35, 2.37, 2.57–58, 3.15C, 3.46B–C, 47A–B, 59–61, 5.2–7).

4.1.7. The Isle of Man Government argued that BNFL's intentions could not be guaranteed in the light of the performance of past Magnox and oxide reprocessing, the latter having culminated in the 1973 Head End accident. They further argued that the data base upon which projections of impact were made was inadequate, and that the basic radiological standards were in doubt. In addition the monitoring programme was under-financed and not extensive enough, and based on a retrospective rather than a future-orientated predictive capability (1.20–22, 93.34–88). This argument was also the opening position of NNC (1.45), TCPA (1.37), WA (1.29), Durham County Council (1.35).

4.1.8. The Lancashire and Western Sea Fisheries Joint Committee (LWSFJC), also held that the present situation was unacceptable, the past control inadequate, especially with regard to radioactive Caesium pollution, and called for stricter monitoring and speedier action on discharge control (2.4).

4.1.9. FOE announced their intention of meeting BNFL's arguments on their own terms, and thus with regard to Health and Safety proffered no critique of the system of control or past record. They concentrated their criticism on the assumption that reprocessing was a proven technology, and on arguing that an alternative was available for the direct disposal of spent radioactive fuel that was at least no worse, and possibly better than reprocessing as far as the environment was concerned, their main argument against reprocessing being the dangers arising from Plutonium as a weapons material (1.22, 1.36).

4.1.10. Parker concluded that for the most part the authorities and the Company could be relied upon to meet the intended targets, that past history bore out this trust, there being only minor matters of procedure and insignificant lapses in competence to which he should draw attention (para. 2.20–33, 10.1–140). However, on matters of comparison, Parker referred to the health hazard from coal, thus assuming that the THORP question was bound up with the future mode of energy production, and concluded that the health risks were greater. This is dealt with in greater detail below as the context of this argument was not re-presented, Spearing having shown that the Ford Mitre Report had assumed no reprocessing, and that reactors could not simply be compared with coal-fired generating stations (see above). Taylor had also emphasised the need for hard evidence on this question (96.16).

4.1.11. Concern about BNFL's intentions and the future impact also centred on the disposal of radioactive wastes as a result of the projected vitrification programme (WA, SCRAM, TCPA and DCC), and the storage of spent fuel (PERG, FOE), as well as the storage on site of liquid waste (PERG, NNC). In the latter case, there was concern expressed also at the hazard potential due to accidents or loss of services to the site (PERG, WA).

4.1.12. In the structuring of the following sections we have distinguished between the problems of risk assessment with regard to the workforce and to the public. Although some of the basic arguments, particularly concerning standards, apply to both there is a division in the responsibilities of the authorities, and also in the monitoring programmes, to which the bulk of the argument referred. We have also distinguished between argument referring to routine (and hence to some degree intentional) discharges and to accidental (and potential) emissions.

4.1.13. Interwoven in the critique of risk assessment were arguments relating to the system of control generally. Although we have maintained a necessary overlap, we have tried to separate some of this general critique into one section in an attempt to relate the technical aspects to matters of acceptability and to the argument concerning democratic accountability which we deal with under another section. We thus, to some extent, make the artificial distinction of risk assessment and institutional safeguards. In the former, we deal with such matters as the history of discharges, the present situation and BNFL's intended future operations, the results of monitoring and the technical disputes regarding the accuracy of the data, the causes of discharges etc.; where the workforce is concerned we deal with the argument concerning the accuracy of the risk assessment for workers' exposure.

4.2. THE BASIS OF CONTROL

4.2.1. The history of the controlling authorities with regard to radiation hazards is complex. At the time of the inquiry almost all areas were either under review, or had recently been re-organised in the light of recommendations made by the RCEP (paras. 527—530, 533). The government's response to the Commission is contained in a White Paper (BNFL 170, especially Annex A, paras. 13—31).

4.2.2. The system as adhered to by BNFL and described by Parker (paras. 10.1—10.17) is as follows: (1) the chief international body concerned with radiological protection standards is the International Commission on Radiological Protection (ICRP). This body recommends radiation limits. Its members are elected by the International Congress of Radiologists.
(2) Other international bodies which have advisory roles are: (i) the United Nations Scientific Committee on the Effects of Atomic Radiation (UNSCEAR); (ii) the IAEA; (iii) the World Health Organisation; (iv) the Nuclear Energy Agency (NEA) of the OECD; (v) the FAO (Food and Agriculture Commission). (3) EURATOM has the power to fix minimum standards throughout the EEC.

4.2.3. How far these international bodies represent a plurality was open to question, many members sitting under different hats; the US system was not described in detail, but does involve additional groups such as the Advisory Committee on the Biological Effects of Ionising Radiations (BEIR), which was used in evidence by Spearing and presented in documentation by BNFL (BNFL 135). The US then has an Environmental Protection Agency which fixes standards, the EPA (JKS 1, 2).

4.2.4. The essential difference between the US and UK systems is that the ICRP recommendations embody a judgement as to acceptable exposure levels, and these are then advised upon by the NRPB and the MRC as to their applicability in the UK. EURATOM standards are also derived from ICRP. The US system differentiates between risk assessment and acceptability, there being separate committees for each, e.g. BEIR and USEPA respectively.

4.2.5. In the White Paper, the government announced that NRPB were to advise on acceptability of the standards of ICRP/EURATOM and the MRC on the biological bases.

4.2.6. The actual fixing of standards with regard to particular activities is complex, with a long history of change and of divided responsibilities. The only Act of Parliament, the 1960 Radioactive Substances Act, stipulates the nature of the authorising process only—that of the joint authorisation of the Secretary of State for the Environment and the Minister for Agriculture Fisheries and Food. In practice this now involves HM Alkali and Clean Air Inspectorate (ACAI) and the Fisheries Research Laboratory, and the operation of a joint committee under the Radiochemical Inspector at the Department of the Environment. The principles of control follow only from a White Paper, of 1959 (BNFL 83), which was under review at the time of the Inquiry. In para. 117 it states the aims for waste disposal: (a) To ensure, irrespective of cost, that no member of the public shall receive more than 0.03 rem weekly, i.e. one tenth of the ICRP level for occupational workers, except (where use of an overall safety factor takes consideration of unusual individuals in a population, e.g. excessive consumers of fish etc.). (b) To ensure, irrespective of cost, that the whole population of the country shall not receive an average dose of more than 1 rem per person in 30 years. (c) To do what is reasonably practicable, having regard to cost and the national importance of the subject, to reduce the doses far below these levels.

4.2.7. Thus the basis in UK law and EEC law rests on the ICRP recommendations of dose limits. The latest ICRP recommendations are contained in Report No. 26 (G 35). Recent changes involve the abandonment of population dose limits which were originally instigated for protection for genetic effects, the view being taken that if individual dose limits were observed, average doses to populations would be within acceptable limits (G 35, paras. 129–130).

4.2.8. The control limits are stated in Mummery's proof. For our purposes the most significant are the limits for members of the public, which are 500 mrem/a, and the workforce, which are 5 rem/a. BNFL's environmental impact intentions were stated in terms of the percentage of ICRP.

4.2.9. A further complication arises from the only other statute, the Nuclear Installations Act of 1965 (PERG 25). Section 7 of the Act states that it shall be the duty of the licensee: (a) To secure that no occurrence involving nuclear matter—on the licensed site, or within the limits in the course of carriage to the site or on behalf of the licensee—causes injury to any person or damage to any property of any person other than the licensee. (b) To secure that no ionising radiation emitted from anything on the site which is not nuclear matter, or primary waste discharged in any form on or from the site, cause injury to any person or damage to any property of any person other than the licensee.

4.2.10. This matter was taken up by Taylor in an attempt to clarify its relevance to waste discharges. Glidewell offered a clarification which essentially stated the problem in terms of liabilities (cf. 3.4.3). Kidwell further took up the problem of claiming damages and the difficulties of proof and Parker referred to it in his report (para. 10.123−124). However, all questioning on the interpretation of the term 'injury' in the act was disallowed. It was thought by Mitchell in answer to a question by Taylor, that the NII could take responsibility for discharges under the terms of the Act but had chosen not to. Taylor concluded that the terms of the Act were too restrictive for the industry and that its basis in law had not been clarified. Research and monitoring had been carried out by a variety of organisations according to the divided responsibilities of MAFF, DOE and NRPB. The RCEP made a number of recommendations to improve the situation, arguing that a unified view was necessary (paras. 528−529). The Government response outlined a number of reviews and the formation of a joint NRPB/MRC committee to co-ordinate research. Otherwise the monitoring programmes would remain with FRL and ACAI, with no unified Pollution Inspectorate as recommended by RCEP.

4.2.11. Details of how the control system operates in practice were given in a series of government documents (G 1, G 2 outline the divisions of responsibility of the government departments; G 10, G 8 on methods of monitoring and dose calculations, with G 19 giving a model dose calculation for caesium in fish and ruthenium in porphyra; G 28 is a draft limitation for future Magnox; G 14 and G 4 the methodology of safety assessment and the role of the NII).

4.2.12. Hermiston gives details of the discharge formulas and statutory limits which are designed to keep below the ICRP exposure limits for each of the pathways back to man (Hermiston paras. 43−52). The present limits for each three month period are: gross beta: 75,000 Ci; ruthenium: 15,000; strontium: 7500; alpha: 2000. The limits under negotiation for the refurbished Magnox plant (G 28) are reproduced below:

4.2.13. The Local Authority (CCC) has no powers of control, and Glidewell submitted that CCC did not wish to have, relying on the appropriate Government authorities (97.74A−D, 32.8−11, 12).

Draft limitations for the authorisation to dispose of radioactive waste from the Windscale pipeline

The limitations are that in all the waste disposed of by this means in any one period of three consecutive calendar months:
- (i) the sum total of curies of Ru-106 does not exceed 5500
- (ii) the sum total of curies of Sr-90 does not exceed 7500
- (iii) the sum total of curies of Zr-95/Ni-95 does not exceed 7500
- (iv) the sum total of curies of Cs-137 does not exceed 10,000
- (v) the sum total of curies of all other beta emitters, excluding tritium and Pu-241 does not exceed 44,000
- (vi) the sum total of curies of tritium does not exceed 25,000
- (vii) the sum total of curies of all alpha emitters together with the alpha activity of Am-241 which will arise from the decay of Pu-241 does not exceed 1000

NOTE: The authorisation will include no additivity formulae.

4.3. BACKGROUND TO RISK ASSESSMENT

4.3.1. The assessment of the health and safety risks associated with reprocessing proved a complex issue. We have separated the evidence and argument into three areas: the background to risk assessment; statements of intent; and institutional safeguards. The first of these categories includes the history of discharges, their source and magnitude, and the past exposures of the public and the workforce. For the most part the data itself was uncontested. Some controversy arose as to the exact source of the discharges. Great controversy arose as to the interpretation of the data in terms of both present health effects and future impact. Most of this controversial area we deal with in section 4.5 as it involves questions relating to the institutions set up to safeguard the workforce and the public. Further controversy surrounded the accident hazard potential of the site and we deal with this below. Lastly, when assessing risks to society there is the complex area of social impact, involving a complex of values relating both to health (in particular mental health) and also to other aspects such as aesthetic qualities of life, moral imperatives, social and political values. We deal with these under 'societal risks'.

THE HISTORY OF DISCHARGES TO THE ENVIRONMENT

4.3.2. The Windscale site was developed by the Ministry of Supply, Atomic Energy Department, shortly after the war, when it had been an Ordnance factory. Two 'piles' were constructed for the production of plutonium for the UK weapons programme, and the first pile went critical in June

1951. These 'elementary reactors' were simple in construction, the uranium being air-cooled in an open circuit, with a discharge through a stack with filters.

4.3.3. In parallel with the construction of the piles a chemical separation plant was built to treat the irradiated fuel and extract the plutonium. This was commissioned in 1951 (known as B.204). After the development of the Magnox gas-cooled reactors for commercial electricity, it was realised in 1958 that a second separation plant would be needed. This was commissioned in 1964 (known as B.205).

4.3.4. Marine discharges began in 1951. It is not clear when monitoring started by the various authorities, but the UKAEA issued monitoring data until 1971 when the site changed names and became BNFL. Investigations of the possible pathways of radioactivity in the environment had been carried out prior to the release of wastes to the sea (BNFL 165).

4.3.5. Aerial discharges began with the commissioning of the plutonium piles, and were added to by the various venting systems of the separation plant, fuel ponds and waste storage silos. The four Magnox reactors, owned by the UKAEA also contribute to the aerial discharges at the site.

4.3.6. Solid waste has been buried at the Drigg site adjacent to Windscale. Scott stated that this material consisted chiefly of slightly contaminated clothing, paper, plastics, and building material.

4.3.7. Thus the environment of Windscale has been subject to discharges for over 25 years. There are also a number of separate facilities on the site: four Magnox power reactors owned by the UKAEA for the production of plutonium, with electricity as a by-product for the site and the grid; one experimental AGR; UKAEA laboratories concerned with aspects of the weapons programme.

4.3.8. The latter services are implicated in discharges but due to security problems, no details were available. It was simply stated that the controls were based on total discharge from the site. This has important implications for the argument concerning control technology for discharges and the history of releases from the site.

Marine Discharges

4.3.9. Hermiston provided a history of the marine discharges (proof para. 69–77, table 1). However, details were provided only from 1971, and prior to that it was stated: 'broadly speaking, discharges to the Irish Sea have increased from about 20,000 Curies per annum (for beta) in the 1950s, to an average level of 70,000 Ci/a until the early 1970s' (para. 70).

4.3.10. The rate of discharge increased to 2000,000 Ci/a of beta activity in the 1970s due to corrosion of Magnox fuel in the ponds (para. 70).

4.3.11. Taylor requested more historical detail and the Inspector ordered this 'subject to security', BNFL supplying a further table (BNFL 208). This and the previous table are reproduced below, tables 1 and 2.

Table 2. Liquid effluent discharges from Windscale 1964−76 (curies per year).

Year	Alpha	Beta	Ru-106	Sr-90	Ce-144	Zr-95	Nb-95	Cs-137	Cs-134*
1964	288	60660	24504	972	3216	21560	20880	2800	
1965	420	55920	20148	1512	3888	17480	32200	2960	
1966	588	65568	24924	912	6852	14080	23360	4890	
1967	960	72264	17232	1392	13704	18800	25720	4050	
1968	1416	84996	24204	1356	9960	28080	37160	10040	
1969	1332	100500	22896	2940	13536	31560	30120	12060	
1970	1668	109716	27660	6276	12480	9080	9920	31170	6775
1971	2688	160200	36468	12332	17252	17380	18120	35820	6372
1972	3858	140632	30500	15160	13564	25624	23520	34840	5815
1973	4894	126932	37800	7444	14548	14900	28100	20770	4481
1974	4572	206584	29160	10648	6532	2560	6996	109770	26993
1975	2310	245148	20556	12636	5608	2629	5924	141377	29211
1976	1614	183482	20698	10344	3996	3099	5980	115926	19953

*Cs-134 not measured separately prior to 1970.

4.3.12. Thompson (for NNC) had followed the history of the discharges in some detail and presented additional tables from symposia (NNC 35, NNC 52). These provided further data back to 1957. We reproduce these below as tables 3 and 4.

Table 3. Discharges from Magnox 1 (NNC 35).

	1957	1958	1959	1960	1961	1962	1963
Ru	26616	42264	35472	39624	25140	22992	33372
Beta	64392	82152	91900	77532	47772	44904	48240
Alpha	57.6	62	67	81	133	186	228

Table 1. Discharge of liquid radioactive waste into the Irish Sea from Windscale and Calder Works 1971–76.

Nature of Radioactivity	Average amount discharged in three consecutive months (curies)						Authorised limit of discharges (per 3 consec. months)	Percentage of authorised limit					
	1971	1972	1973	1974	1975	1976		1971	1972	1973	1974	1975	1976
Beta activity	39180	35483	31883	48714	62313	46821	75000 curies	54	47	43	65	83	62
Strontium-90	3094	3829	1925	2465	3285	2608	7500 curies	41	51	26	33	44	35
Ruthenium-106	9100	7600	9246	7542	5276	5162	15000 curies	61	51	62	50	35	34
Cerium-144	4485	3415	3584	1698	1593	1009	—						
Alpha activity	661	930	1182	1173	657	382	2000 curies	33	47	59	59	33	19
Alpha activity discharged in 12 months:	2688	3858	4894	4572	2310	1614	6000 curies	45	64	82	76	39	27
Additivity formula	0.79	0.66	0.76	0.68	0.58	0.51	1.0	79	66	76	68	58	51

Note: Discharges through the pipeline also include tritium. Current authorisations do not specify a limit on tritium, which has negligible radiological significance, but the discharges represent less than 0.1% of the ICRP limit.

Table 4. Ru-106, Zr-95 and Nb-95 discharges from Windscale (NNC 52). All discharges in Ci/year.

		Ru-106	Zr-95	Nb-95
(treatment plant)	1960	39600	2400	6300
	1961	25300	1700	7900
	1962	23000	940	4300
	1963	33400	560	3300
Magnox 2	1964	23100	21600	20800
	1965	21000	17700	33700
	1966	24900	14100	23400
	1967	17200	18800	25700
	1968	24200	28100	37100
Zr/Nb reduction action	1969	22900	31600	30000
	1970	27600	9100	9900
	1971	36400	18000	17300

4.3.13.   BNFL had provided a document by Dunster *et al.* to which little reference was made, which gave an account of the experimental period of marine discharges between 1952 and 1953. This document (BNFL 165) also gave further details of discharges and we reproduce these in table 5 below.

Table 5(a). Effluent discharges, 1953–62.

Period	Mean monthly discharge rate (Ci/calendar month)			
	Total alpha	Total beta	Ru-106	Sr-90
Apl 1953– Feb 1956	4.3	1720	510	50
Mar 1956– Dec 1957	5.1	6140	710	156
Jan 1958– Dec 1960	5.8	7310	3260	127
Jan 1961– Dec 1962	13.3	3860	2010	63

Table 5(b). Mean discharge rate of beta emitters.

Nuclide	Discharge rate (Ci/calendar month)			
	1959	1960	1961	1962
Ru-106	2956	3302	2095	1916
Ru-103	746	964	266*	153
Ce-144	583	74	180	200
Cs-137	165	76	92	92
Zr-95	845	196	140*	78
Nb-95	415	481	658*	—†
Y + R.Es.‡	506	83	201*	125
Sr-89	170	82	114*	42
Sr-90	129	43	42	85
Total	6515	5301	3788	2691
Total beta activity (Sr-90/Y-90 standard)	7659	6461	3981	3742

* Average for first 8 months of the year.
† Not available.
‡ Rare earths.

4.3.14. In addition Urquhart requested details of Cs-134 (BNFL 181, incorporated in BNFL 208), and curium (BNFL 218). BNFL also provided a breakdown of plutonium and total alpha since 1972 (BNFL 220). Curium varied from 4 Ci/a in 1965, to a maximum of 49 in 1973, with 16 in 1976.

Table 6. Plutonium discharges and total alpha.

	Pu	Total alpha
1972	1547 Ci/a	3858
1973	1768	4894
1974	1248	4572
1975	1111	2310
1976	1266	1614

Table 7. Alpha activity discharges.

Year	Total alpha activity	Pu-238, 239, 240	Am-241	Remainder mainly curium and a trace of neptunium and almost negligible quantities of Pu-242	
1968	1416	828	576	12	1%
1969	1356	816	396	144	10%
1970	1656	936	540	180	11%
1971	2688	1128	1020	540	20%
1972	3864	1548	2172	144	4%
1973	4896	1776	2952	168	3%
1974	4572	1248	3192	132	3%

4.3.15. Urquhart also provided an amplification of data given in IOM 82, for alpha activity (WIERC 49) (table 7).

4.3.16. Thompson (for NNC), provided a further breakdown of alpha activity discharges in relation to the throughput of fuel in the separation plant (cf. para. 4.3.43).

4.3.17. Warner additionally provided some figures for tritium and iodine-129 releases from the Magnox plant: H-3 less than 40.000 Ci/a (para. 147, Appendix 1); I-129 approx. 6 Ci/a (para. 148, Appendix 1).

Aerial Discharges

4.3.18. The sources of aerial discharges from the Magnox plant are given in Warner (Proof Appendix 2, especially paras. 50–51, 138, 178–182). The methods of containment and decontamination factors are described. Hermiston (Proof table 3) provides the history of discharges since 1971. There was no detailed questioning on sources or control of discharges to the atmosphere with respect to the past record of Magnox.

4.3.19. Table 8 is reproduced from Hermiston.

4.3.20. Additionally, BNFL reported that discharges of caesium-137 and strontium-90 had risen in recent months due to problems with the solid waste silos which were the chief source of these nuclides. The discharges were due to activity entrained in the ventilating air. The Company reported that recent discharges had led to an increase in the contamination of milk in the adjacent farms. The 1976 levels had been 2.4% of the limit for strontium and 0.2% of the limit for caesium, whereas levels had now risen to 5% and 1% respectively. Measures to control this were being taken (BNFL 318).

Table 8. Airborne radioactive waste discharges to atmosphere, 1971–76.

Source and nature of radioactivity	Average discharge in three months (curies)					
	1971	1972	1973	1974	1975	1976
Beta activity	5.7	0.77	4.75	0.70	0.48	0.85
Alpha activity	0.11	0.034	0.045	0.045	0.019	0.013
Strontium -90	0.21	0.027	0.18	0.037	0.051	0.048
Iodine -131	0.14	5.5 (2)	0.30	0.003	0.0023	0.019
Krypton-85	3×10^5	3×10^5	2×10^5	2×10^5	3×10^5	3×10^5
Tritium (1)	3×10^3	3×10^3	2×10^3	2×10^3	3×10^3	3×10^3
Argon -41	10000	10000	10000	10000	7500	5400

The table contains a summary of the discharge from the main stacks at the Windscale and Calder Works. In addition, since 1974 there has been an increase in low-level discharges on the site.

The average three monthly discharge rates in the last two years have been: beta activity 0.3 curie; strontium 0.05 curie; caesium -137 0.20 curie; alpha activity 0.05 millicurie.

(1) Tritium values are inferred by comparison with krypton -85.
(2) This includes an exceptional discharge of 20 curie I-131 in December, caused by the inadvertent feeding of short cooled fuel elements to the Separation Plant.

STORAGE AND DISPOSAL

4.3.21.   In addition to gaseous and liquid wastes discharged to the environment, there are several forms of radioactive wastes varying from liquid to solid with intermediate sludges, which are variously termed Highly Active (HA), Medium Active (MA) and Low Active (LA). Table 21 (para. 4.4.6) illustrates the various categories of liquid waste. At present LA and some MA liquid streams are discharged to the environment, and LA solids are disposed to burial at Drigg. Some MA liquid is stored until sufficiently decayed to be discharged under the authorised limits. However, a great deal of solids, sludges and HA liquors are stored on site and await the development of disposal techniques that will prove acceptable (cf. 4.5.93–97, 100–103, 123–129).

4.3.22.   We detail here the source, rate of arisings and present methods of storage of the various waste categories. In a later section we deal with the controversy surrounding future intentions (4.5.123–129).

High Active Liquid Waste

4.3.23. The HA effluent streams (see table 21) of a volume of 24 m^3 per day, contain 1 MCi of total beta and 2000 Ci of alpha activity (per day). These streams are evaporated to leave a concentrate for storage (Warner, Appendix 2 paras. 109–128). The first Magnox plant fed stainless steel tanks of 70 m^3 capacity, of which there are now eight. The second plant fed tanks of 150 m^3 capacity, of a new design (paras. 129–137), first commissioned in 1970. There are now four in use, one acting as a spare, and further tanks are planned at the rate of one per year, with every fourth tank acting as spare. At the time of the Inquiry one was nearing completion for use, and one as spare.

4.3.24. The rate of arisings measured at the entry point to the tanks depends on the performance of the evaporators and the nature of the feed which depends on the burn-up of the fuel. Recent concentrates have varied from 33 l/te uranium processed to 55 l/te with higher burn-up (para. 134).

4.3.25. The tanks have a design lifetime of 50 years, the limiting factor being the corrosion resistance of the internal components. The intention for disposal of Magnox HAW is to glassify the liquids and to bury the glass cannisters (cf. 4.5.93–97).

High Active Solid Waste

4.3.26. The fuel element cladding is both highly active and contaminated by fission products including alpha emitters. Clelland characterised this as MA, but agreed with Taylor that it constituted HA waste and was so defined internationally (17.74). Clelland gave a figure of 0.25–0.6 m^3 per te of fuel reprocessed as the arisings depending on fuel type. At present Magnox elements are stored underwater (they are inflammable in air) in underground silos (Clelland proof para. 12).

4.3.27. Taylor drew attention to the figures of the RCEP (BNFL 9, table 9, p. 139) on the arisings of solid waste and the technical problems of waste silos (cf. BNFL 318). These figures are reproduced in table 9.

Medium Active Wastes

4.3.28. Apart from the MA effluents, there are various sludges and ion exchange resins from the Magnox process, and contaminated graphite sleeves from the small amount of AGR fuel processed. Clelland outlines the present position and intention (proof, paras. 14–15). Sludges can be fluidised easily and solidified for future disposal and the graphite will be packaged.

Table 9. Volumes and activities of various types of solid waste accumulated at Windscale in 1974 and estimated for 1985. The comparable figures for HA liquid waste concentrate are also given.

Type of Waste	Volume, m³		α-activity, kCi		β-activity, MCi	
	1974	1985	1974	1985	1974	1985
Fuel element cladding	4000	13000	10	100	1	15
Sludge from treatment of liquid waste (obsolete process)	5000	5000	10	10	–	–
Plutonium-contaminated (low βγ content)	3000	5000	60	100	–	–
High-level liquids	550	1800	500	2000	350	4500

Low Active Wastes

4.3.29. Apart from the LA effluents, a variety of wastes of some bulk and in particular plutonium contaminated waste low in beta and gamma activity, have arisen (Clelland paras. 18–21). Some of this waste is packaged and buried at Drigg, (Hermiston proof table 4, para. 70). The history of the Drigg site is given, and over 80,000 m³ had accumulated between 1960 and 1970. Recent rates of disposal were over 8000 m³ per annum. Formerly Windscale accounted for 50% and the UKAEA 40%, but in recent years Windscale's percentage had risen to 80%. Approximately 1500 Ci/a of beta emitters, and 10 Ci/a of alpha activity was disposed of in this way. Volumes disposed of, taken from Hermiston's table 4, are given in table 10.

4.3.30. Clelland (para. 20) stated that much of the miscellaneous plutonium contaminated waste could not be disposed of to Drigg as the activity content was too high. This waste is stored and awaiting decontamination, packaging and eventual disposal.

Issue: Whether the past record of discharges was indicative of problems with the technology of reprocessing, fuel storage, or discharge control.

4.3.31. The record of discharges as presented in the documents of the Company and of the authorising bodies was not disputed. However, controversy arose over the ascribed cause of the discharges. BNFL maintained that

Table 10. Solid waste disposals at Drigg.

Source of waste	Volume (m³)					
	1971	1972	1973	1974	1975	1976
BNFL Sites						
Windscale/Calder	6763	8890	11700	19454	8124	11728
Springfields	79	72	43	58	39	710
Chapelcross	161	84	82	83	56	69
Capenhurst	2	16	7	1	-	7
Nuclear Establishments						
AERE, Harwell	566	488	405	465	848	811
AWRE, Aldermaston	357	295	426	370	164	373
AEE, Winfrith	302	337	240	428	498	545
REML, Culcheth	-	-	30	8	-	12
RCC, Amersham	76	79	145	324	222	548
SSEB, Hunterston	228	218	250	292	213	454
National disposal service	140	779	460	1504	663	449
Overall total:	8674	11258	13788	22987	10827	15706

the technology of reprocessing was well proven; that the THORP process would involve differences in degree only from previous experience with Magnox reprocessing, and that consequently they were able to make an appropriate statement of intent with regard to future discharges (cf. 4.4).

4.3.32. The TCPA, NNC and PERG analysed the past record of discharges and argued that it was indicative of technical problems that had not been foreseen. The question of the adequacy of the response to technical problems (some of which were admitted by BNFL in evidence), is dealt with under Insitutional Safeguards. The controversy we deal with here is that arising from the technical reasons given by BNFL for the discharges.

4.3.33. Thompson for NNC provided the main technical evaluation of BNFL's discharge record, with Urquhart questioning in considerable detail. Their main arguments concerned the technical problems of controlling caesium, zirconium/niobium, ruthenium and alpha emitters.

Caesium

4.3.34. BNFL acknowledged that there had been a rapid and unforeseen increase in the discharges of the two radio-isotopes of caesium, CS-137 and 134. This had led to unacceptably high exposures to local fish consumers and to an increasing population dose to fish eating people in both the UK and Europe. It was emphasised by BNFL that the discharges were at all times within the authorisations and that no person had received in excess of the recommended dose limits, nevertheless the situation was unsatisfactory (Hermiston paras. 69–77, Allday 3.42). This situation was documented graphically by CCC as illustrated in figures 4 and 5.

4.3.35. There was some dispute as to the cause of the caesium discharges. BNFL maintained that the rise was almost entirely due to problems arising in the fuel storage ponds. Hermiston and Warner (proof, Appendix 2 paras. 32–40) outlined the technical problems: Magnox fuel cannot be stored under water for periods in excess of 6 months; this necessitates a constant rate of reprocessing at close to capacity to deal with the arisings from UK reactors; due to the three-day-week in 1973/74 there was a hold-up in the installation of high active storage tanks and the Magnox plant which fed wastes into the tanks had to be closed down; the backlog of fuel began to corrode and corroded fuel became difficult to shear for dissolution; the water in the fuel ponds thus became contaminated and affected the working space leading to unacceptably high doses to the workforce; the ponds were not equipped with effluent treatment plant and thus the contamination had to be discharged to the Irish Sea direct.

4.3.36. Measures were taken to control the discharges by the use of Zeolite 'skips' which absorbed the caesium in the pond water. In addition Warner indicated that a new pond-water treatment plant would be operative by 1980/81 and this would further reduce the effluent.

4.3.37. Thompson (NNC) maintained that the Company already had problems with caesium due to the increasing caesium content of Magnox fuels which had been irradiated for much longer periods than anticipated. There had been a five-fold increase in the radioactivity of fuel for reprocessing (67.34C–35). However, Thompson and Urquhart maintained that there was evidence to implicate the storage of oxide fuel in the releases of caesium. (i) The amount of Cs-134 discharged, characteristically present in the longer burn-up fuels in much greater amounts than Magnox, increased 6-fold in 1974 when oxide fuel was both reprocessed and stored in large amounts. (ii) The ratio of Cs-137/134 altered in the discharges during this period (WIERC 5). (iii) The ratio of americium to plutonium also altered during this period (WIERC 7). (iv) There was evidence of problems with foreign fuel, particularly Japanese, in that the elements were showing a high rate of failure in the reactor and in transport (WIERC 1). (v) BNFL had already insisted that all new arrivals of oxide fuel be bottled before despatch (9.39–47).

4.3.38. Warner had stated in answer to Urquhart that the source of caesium pollution was primarily the ponds and not the separation plant. He

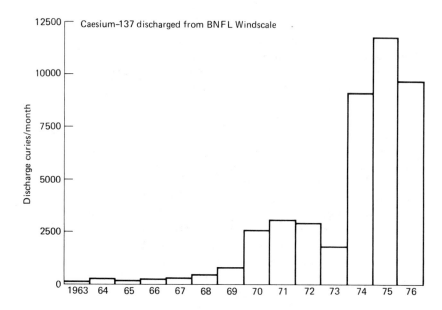

Figure 4. Caesium–137 discharges.

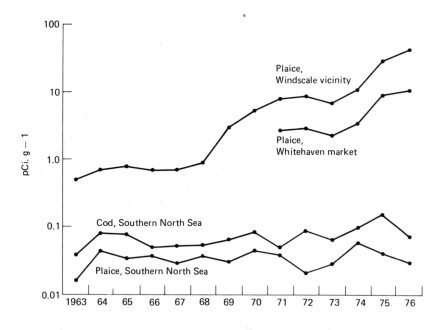

Figure 5. Caesium–137 in fish flesh from the Irish Sea and Southern North Sea.

maintained that the separation plant discharged approximately 44,000 curies of beta per annum, and the ponds were discharging in excess of 140,000 Ci of beta. Furthermore, the discharges were not related to the throughput of the separation plant in a simple manner, but were a function of the effluent tank-management programme (29.34).

4.3.39. BNFL produced evidence of the discharges of caesium according to types of fuel stored (BNFL 202, 225). These documents showed that in 1973 oxide effluent accounted for 430 Ci/annum, and in 1974, 280 Ci/annum, whereas Magnox figures were 18,000 and 131,000 respectively. Thompson refused to believe these tables, arguing that they did not account for the 'co-incidences' he had documented (68.36A–39C).

Zirconium/Niobium

4.3.40. Thompson provided evidence that the releases of Zr-95/Nb-95 which had risen following the changes brought by the introduction of the second separation plant had not been expected (67.34F–G, NNC 72). These discharges had led to an exposure pathway as the radionuclides accumulated in the silt of the Ravenglass estuary and irradiated the salmon fishermen who tended a garth (see also table 4).

Ruthenium

4.3.41. Thompson provided evidence that Ruthenium control had been a long-standing problem. It had accounted for the major exposure pathway arising from the early reprocessing of Magnox fuel, as it accumulated on the sea-weed Porphyra, which had been at that time an export commodity to communities in South Wales who made it into laverbread (cf. 4.5.22–26, 4.5.77–84). The highest ruthenium levels had been in 1958 (see table 5), and these were brought under control by the installation of an effluent treatment plant in 1961 (67.23D–F–24B–C). Predictions were made with regard to the second separation plant (NNC 35 and WIERC 46), and in 1964 there was a reduction in ruthenium discharges. However, by 1971 discharges were back to the highest of previous levels (67.26D).

Alpha Emitters

4.3.42. Thompson and Urquhart sought to identify the nature of the in-creased alpha-discharges—plutonium-238, 239, 240, americium-241 and curium. In addition plutonium-241 decays with a half-life of 13.7 years, and Thompson pointed out that the daughter was the alpha-emitting americium 241, and that in 1974, 35,000 Ci had been released (para. 88). Silsoe elaborated on the last point, in that it would take 80 years for the decay to produce 1200 Ci of alpha (68.43B–G).

4.3.43. Both Thompson and Urquhart suspected that the increase in alpha (the authorisation had to be renegotiated in 1970 to allow the increase, cf. 4.5.115–122), was due to problems in the separation plant or fuel ponds that were themselves a result of the input of oxide fuel. Thompson illustrated the relationship in Table 12 of his proof (see table 11).

Table 11. Actinide discharges from Windscale.

Year	Magnox proc.d. te	Oxide proc.d. te	Total alpha Ci	Pu -238,9,40 Pu Ci	Am -241 Ci	Am/Pu	Cm Ci
1968			1416	828	576	0.70	14
1969		23.3	1356	816	395	0.49	13
1970	1177	0.0	1656	936	540	0.58	17
1971	1086	21.1	2688	1128	1020	0.90	27
1972	765	41.8	3864	1548	2172	1.40	39
1973	730	5.2	4896	1776	2952	1.66	49
1974	1121		4572	1248	3192	2.56	46
1975	589		2310	1111	1176	as alpha - Pu, Cm.	23
1976			1614	1266	332		16

4.3.44. However, Warner under cross-examination by Urquhart, revealed that at least two-thirds of the plutonium effluent came from 'other activities' on the site (29.34G–H and 35H). When pressed, Warner stated that he could give no further details for security reasons, except to say that the separation plant put out approximately 500 Ci/annum of alpha-emitters, and that no americium came from the reprocessing activities because this was water soluble and hence separated out into the High Active stream going into the HA waste storage tanks (30.15C). Such americium as was discharged came from 'other activities' (30.16A). At one point these activities were described as 'plutonium recoveries' (30.14D). They were also anticipated, according to Warner, 'from a load of work not connected with the separation plant' (29.36G). He also referred when asked to compare Windscale discharges with those from La Hague to the French 'special recovery' processes as not being situated at La Hague (30.17F and 18).

4.3.45. These activities were alluded to in Warner's appendix 2, at para. 11, where he describes an annex to B.204 being used prior to 1964 as a plutonium purification plant and then post 1964 for the recovery of 'aged plutonium residues'. From the evidence above it can be concluded that these special services have military significance (Windscale is both a civil installation and a defence installation), and that the same form of effluent treatment is not available to the B.204 annex (see also Author's note below).

4.3.46. When asked to compare the 500 Ci/a release of alpha-emitters at Windscale, to the French authorised limit of 90 Ci/a and actual discharges of 27 Ci/a (cf. 4.5.115–122, 4.5.93–97), Warner pointed out that the French plant had a flocculation treatment for the effluent, and that this had been contemplated for Windscale's second separation plant, but was not installed—this accounting for the discrepancy pointed out by Urquhart in the expectations of Williams and Davidge (WIERC 46) and the actual discharges (30.13B–F).

Case references
Allday 9.39–47; Thompson 67.23–35 and 68.36–43; Warner 29.34–36 and 30.13–18

Proofs
Hermiston; Thompson (NNC); Warner

Documents
BNFL 165, 181, 208, 218, 220, 225; CCC 42, IOM 82; NNC 35, 52, 72; WIERC 1, 5, 7, 46, 49

THE PAST RADIATION EXPOSURE OF THE WORKFORCE DUE TO MAGNOX REPROCESSING

Issue: Whether the past exposure of the Windscale workforce had led to significant health effects, and whether there was evidence that current assessments of risk were accurate enough to justify planning permission.

Author's note: Plutonium is the constituent of nuclear warheads. As the warhead ages the decay-products poison the system, making it less stable and predictable. Hence, warheads must be periodically re-processed to sufficient purity.

4.3.47. In this section we outline the past history of exposure to the workforce and the health statistics that were put in evidence. According to the schema we have adopted, the interpretation of health statistics and the questioning of basic standards is dealt with in section 4.5 under 'institutional safeguards'. However, we here introduce the arguments in relation to risk assessment for the workforce. Various points are followed up as they relate to the authorising bodies in section 4.5 (cf. 4.5.18–21, 49–60, 61–65).

4.3.48. BNFL gave evidence to show that the Windscale workforce was approximately 20% healthier than equivalent industrial workforces, except for ischaemic heart disease, where it was 20% above the expected. All cancers were below expected (see tables 12 and 13). Mummery referred to the NRPB survey of cancers of the workforce (BNFL 119) which had found no correlation of cancers with radiation doses received. Schofield presented the companies medical procedures, screening and an outline of the record. Mummery outlined the system of standard setting and control and the company's record of meeting the regulations: the average dose to the workforce was 1.2 rem per man per annum, and no person had ever exceeded the statutory life-time dose of $5(N-18)$, where N = age in years. However, he admitted that 'quite substantial' numbers of workers had exceeded the ICRP recommended limit of 5 rem per annum. In answer to Barratt (19.36), Mummery gave a breakdown of the percentages exposed for the period 1971–76 (20.4), which varied:

71–74% of the workforce were below 1.5 rem/annum
12–17% of the workforce were below 1.5–3
12–14% of the workforce were above 3
02–5% of the workforce were above 5

4.3.49. Since 1971, only six workers had received in excess of 20% of the maximum permissible body burden of plutonium. The lowest figures for external doses above 5 rems had been achieved in the last two years. Hermiston replied to Pochin that this reduction had been achieved by spreading the dose across more workers, hence producing a greater collective dose (27.9–10). In answer to Taylor, Mummery agreed that the total collective dose to the workforce averaged 4000 manrems, and might total 80,000 manrems in a 20 year period, giving on current ICRP estimates of 1 cancer death per 10,000 manrems, some eight radiation induced cancers in the workforce (20.74F–76D).

4.3.50. This evidence was not disputed, however questions were raised as to the comparability of the Windscale workforce with other industries for purposes of calculating the expected frequency of cancers. Mummery, cross-examined by Taylor, acknowledged that the Windscale workforce could not be readily compared because of the selection procedures (20.79–82). However, Schofield did not feel the workforce was highly selected in spite of the company's medical vetting procedures (22.13C–F). Dolphin agreed the prob-

Table 12. Comparison of actual to expected deaths Windscale and Calder Works: male employees and pensioners 1962–75 inclusive.

Cause of death	ICD No	Pensioners aged under 65 and employees		Pensioners aged 65 and over		All employees and pensioners		Ratio A : E
		Actual	Expected	Actual	Expected	Actual A	Expected E	
All causes		296	360	150	184	446	544	0.82
All neoplasms	140–239	72	97	33	43	105	140	0.75
Diseases of circulatory system	390–458	176	157	81	89	257	246	1.04
(Ischaemic heart disease)	(410–414)	(138)	(113)	(60)	(54)	(198)	(167)	(1.20)
Diseases of respiratory system	460–519	11	37	14	31	25	68	0.4
Diseases of digestive system	520–577	3	9	6	4	9	13	0.7
Accidents and violence	E800–999	19	27	2	3	21	30	0.7
Others		15	33	14	14	29	47	0.6

Table 13. Actual and expected deaths among males in Cumbria and West Cumbria, 1971–75 inclusive.

Cause of death	ICD No	Age	Cumbria			West Cumbria		
			Actual	Expected	Ratio A : E	Actual	Expected	Ratio A : E
All causes		15 to 65	4683	4485	1.04	1446	1265	1.14
		Over 65	10564	10015	1.05	2811	2525	1.11
All neoplasms	140-239	15 to 65	1145	1215	0.94	341	345	0.99
		Over 65	1958	2050	0.96	506	515	0.98
Ischaemic heart disease	410-414	15 to 65	1770	1530	1.16	595	455	1.31
		Over 65	3522	2895	1.22	939	730	1.29

and Liverpool and were due to lung-cancer caused by inhalation from the passing cloud. In addition Thompson estimated that vast tracts of land would remain uninhabitable due to caesium contamination, as for example indicated in table 13 of his proof, where dose in rads 1 m above ground from ground deposition was in excess of 1 rad, 1 day after release as far away as 90 km.

4.3.96. Thompson's evidence was criticised by BNFL as not being a realistic assessment. It was argued that there were no circumstances under which simple counter-measures could not be taken, such as the use of fire-hoses. Although Thompson readily agreed that his study was not 'realistic' in the sense that sufficient data on the THORP proposal or present tanks had not been provided so that an accurate assessment could be made, he maintained that the initiating mechanisms, which might include sabotage, aircraft impact, severe disruption of supplies by industrial action, or terrorist attack, or an accident synergism caused by spillage leading to evacuation of the site, could all be regarded as realistic and that a prolonged loss of cooling should not be ruled out—particularly as the object of a safety study was twofold: (i) to identify possible alternative designs (he maintained that the apparently high risk from spent fuel could be easily avoided); (ii) to present to the public the full range of possibilities so that they might take their own view as to the acceptable risk.

4.3.97. Referring to this evidence Parker noted, 'Dr Thompson accepted that the various exercises which he had carried out were not intended to be predictions of risk under realistic conditions' (para. 11.17). And, considering that Thompson accepted the view that it would be 'likely to be possible to build THORP to acceptably safe limits', he stated, 'I find it unnecessary to consider a series of other releases (than Donoghue's) which were suggested' (para. 11.16—17). He did not represent Thompson's detailed criteria for acceptability, which included a full and detailed safety study, nor did he present the context for Thompson's acceptance of his study not being a 'realistic prediction of risk' (see 90.1—4).

4.3.98. Taylor concluded for PERG that an open and independent safety study would be a prerequisite for the 'acceptability' of THORP. He also noted that glassification would not solve the problem of HAW tanks for the life of the industry, for either the liquid must be precooled in tanks for decades before glassification, or if vitrified immediately from reprocessing, then the glass would require active cooling in the same way as the liquid, with the potential for melt-down. Taylor also noted that the Nuclear Waste Advisory Committee had recommended changes to the design of tanks, and it was not clear whether these changes had been incorporated into the older designs, nor whether they could be (BNFL 229, and 96.19—21). Furthermore the tanks contradicted a fundamental principle of engineering safety, as stated by the NII in evidence (G 4, p. 4): they did not fail-safe on loss of services such as power or water or personnel, and he noted that Charlesworth had been unable to think of a parallel, and had therefore agreed they were unique in this respect (45.47).

Case references
BNFL 23.59−61, 26.33−36, 27.4 and 90.102−103; Carr 47.74−77; Charlesworth 45.1E−F, 10−16, 22−26, 41−42, 49, 62−63, 67; Clelland 17.57; Corbet 28.26−29; Donoghue 23.25−40, 24.1−46, and 25.1−49; Dudman 26.44−45, Ellis 77.71G; Farmer 52.50−56, 58−64 and 68; Fremlin 36.37−43 and 56−59; Milne 27.10−60; PERG 96.19−21; Robertson 48.31G; Shortis 13.5; Thompson 89.89−109 and 90.1−4; Wakstein 84.96−106 and 85.3−33; Warner 15.60; Wilson 48.9−21

Proofs
Donoghue; Milne; Shortis; Thompson (PERG); Wakstein; Wilson

Documents
BNFL 3, 86, 240, 242, 243, 280, 282, 284, 289, 299, 303, 308, 309; G 4; PERG 9, 12, 13, 55, 71, 72; WA 37, 189

SOCIETAL RISKS

Issue: Whether the THORP decision was likely to lead to adverse social and political effects such as to justify the withholding of planning permission.

4.3.99. Evidence put forward by PERG (Taylor Proof and 96.4−29) was held to indicate that the THORP decision was likely to lead to increased polarisation of the nuclear debate at a national level, with a potential for conflict, and to social health effects at a local level. THORP, it was argued, was symbolic of nuclear expansion, and nuclear technology bred both conflict and anxiety. This was manifested at a local level by doubts and fears leading to mental ill-health.

4.3.100. Taylor referred to recent research commissioned by the IAEA (PERG 32, 33) and to several polls and surveys on the perceptions of nuclear power. The role of scientific controversy, information programmes, debates and argument was outlined. It was argued that in all those countries where more information had been made available, the existence of controversy and doubt had led to an increased rejection of nuclear power. Such evidence as was available in Britain (PERG 35, 39, 40) suggested a similar situation. Taylor argued that the public view of nuclear technology might well manifest itself in apparently 'irrational' and simplistic views, and to focus upon apparently trivial leaks, but what was at issue was the public image of nuclear technology. This image was necessarily different from that of an informed expert, but in some ways it could be regarded as more perceptive. For example, the public view of a leak might embody a perspective more concerned with the failure of 'containment' and the implications of small human errors for the likelihood of future more disastrous failures, than with the actual degree of radiological harm occasioned by the particular leak in question (90.54–55).

4.3.101. There was also a question of psychological damage to communities subject to doubts and fears surrounding a particular installation. Taylor quoted from Pahner (PERG 33), who reported to the IAEA that the public perception of risk was concerned more with the 'mode of death' than with the probability. Nuclear power was viewed as a 'life repressing technology', and the modes of death from potential accidents, involving cancer and genetic damage, constituted a psychological burden to communities exposed to these risks. Pahner concluded, 'if we do violence to these inbuilt values, we disorder our lives, as persons, as groups, as nations and as a world of human beings'.

4.3.102. This aspect of the social impact of the THORP decision was also taken seriously by BCC (70.58), FOE (93.7), SEI (86.43) and SF (88.16). FOE in particular argued that the THORP decision had implications for the public perception of nuclear risks at an international level and referred to a statement by Kistiakowsky, that a decision to build THORP would signal a legitimisation of plutonium recycle with its threat to humanity through the risks of weapons proliferation (FOE 134).

4.3.103. Many local witnesses, both independently, and as part of objecting groups such as FOE-WC, NNC and the BCC, gave voice to the feelings and perceptions with regard to Windscale. These served to underline the analysis of PERG, Taylor argued. In particular Postlethwaite, the Vicar of Whitehaven and for nine years formerly of Seascale, gave evidence that the workforce was under considerable strain because of the anxieties felt by those who did not work there. There was low morale and concern about safety (70.79–93). He had voiced his concern to the NRPB who, it was noted, had written to BNFL stating that they 'do not think he [Postlethwaite] will cause you any difficulty' (70.82). This had destroyed Postlethwaite's confidence in a source of independent advice (70.82F).

4.3.104. FOE-WC put forward several witnesses who testified to the doubts and fears that were prevalent in the local community. Haworth argued that the opposition was not ill-informed, and was most concerned about the dangers of proliferation. It was not convinced the risks needed to be run in order to fill an 'energy gap' which may or may not arise (60.1–7, 61–92). Although he personally was less worried about safety, this was a recurrent concern of the local witnesses, and these fears were compounded by rumours, distrust, past poor public relations, dependency on the works, and inadequate representation on the Local Liaison Committee. Norman, a local farmer, gave an account of the loss of confidence since the Windscale fire in 1957 when large quantities of milk had to be disposed of because of contamination by radioactive iodine (60.18–22). Bainbridge, an employee of BNFL, expressed his doubts on safety and talked of soul-searching in view of the hardening attitudes within the Company and the subsequent suppression of caution. When asked by Bartlett if he had considered leaving, he replied that he could not afford to (60.55).

4.3.105. Corkhill, a nurse in the area for 26 years, told of the psychological stresses, particularly of wives of workers. She was concerned also that the Community Health Council was not represented on the Local Liaison Committee and that the Medical Officer of Health, who was a member, was out-of-touch with the feelings of ordinary people (60.7–16).

4.3.106. Dixon, a local councillor, gave evidence of the political pressures brought to bear on local councils (60.40–52). Jones, Mcleod, and Henderson voiced further doubts on safety, the ability of scientists to contain the wastes, the effects of low-level radiation etc. (60.39, 31–32, 105–106). Dudman also provided accounts of local feelings with regard to safety and health (20.64, 70, 67.68, 24.44). Higham, a member of the Local Liaison Committee and former employee of the AEA at Windscale, maintained that locals were not free to voice objection or concern about safety. She was concerned that 'operator error' could never be controlled. She also noted that the National Federation of Womens Institutes was opposed to the development of the FBR and was particularly concerned about the waste problem (71.1–8).

4.3.107. Halliday, for the Cumbria Naturalist's Trust, stated the Trust's opposition based on a concern for the natural environment, in particular, a possible threat to small local populations of rare species (70.94).

4.3.108. Further local views were presented, as well as from individuals further afield, in which safety and health were important aspects. Recurrent themes were the existence of uncertainty, of disputes and argument among experts, of responsibility to future generations and of an unproven need to take the risks inherent in the nuclear options (Chivall, Fish, Hatton, Holden, Hillier-Fry, Jones, Miller, Paulin, Richardson, Spearing, Sly, Tremlett, Wadsworth and Wynne).

4.3.109. The IOM Government reported 'great public disquiet' on the island (71.31); that there had been an unprecedented petition; and that although their case was concerned with improvements to monitoring and control of discharges, and not with opposition to the plant as such, they would prefer that Windscale did not exist at all (71.48).

4.3.110. Evidence and submissions relating to the broader national and international social impact necessarily merged into the various critiques presented by the interest-groups represented at the Inquiry. In this respect the opposition of such parties as the Conservation Society, Friends of the Earth, the Society of Friends, the British Council of Churches, the Society for Environmental Improvement etc., is evidence itself of the social impact. This was alluded to by PERG, and by Hall, for the TCPA (80.89—91). Goldsmith, for the WA, in particular, drew a picture of the environmental reasons for the opposition of such groups as the Conservation Society, pointing to the social impact of economic growth, the potential instabilities of industrial societies, and to the alternatives that existed (85.33—62). Spencer, for WA, referred to the institutional inertia within the system, and to what he regarded as the insoluble problems of waste disposal and containment (86.20). Wider political considerations were raised by Coates with respect to EEC legislation (64.90—98), by Dalton concerning waste disposal in Australia (96.1—2, 97.1—9), and by a Swiss group concerned with the implications of contracts with the UK for reprocessing and waste disposal (86.94, 87.1).

4.3.111. The viewpoint that the THORP development would be of benefit to the local community, expressed by Copeland Borough Council and CCC, related primarily to the economic impact (see Conventional Planning Issues). The broader social and political benefits were put forward by BNFL, but also stated, by the TUC (TUC 2, 3), by Ridgeway Consultants (Bowie, Fletcher, Greenhalgh, Macdonald, Rippon), and by a number of persons objecting on other grounds, including e.g. Radford, and Bowen, who, although not necessarily convinced of the necessity of THORP, were nevertheless of the opinion that nuclear power was necessary.

4.3.112. In addition, the social impact was further complicated by the doubt and distrust between protagonists. This was referred to in proof by Taylor, and stated in evidence by Little, who regarded the environmental campaigns as a Soviet-inspired plot (47.79—82).

4.3.113. Bartlett pointed out that there was a contradiction in any opposition to THORP alone, when if the present nuclear programme was accepted, then wastes would accumulate and something would have to be done. Logically a party would have to oppose all nuclear power. This was in response to the position taken by BCC (70.67). At various points BNFL made the point that other energy options, particularly coal, presented comparable health hazards from routine emissions and that it could also be considered to be in the interests of future generations to use nuclear fuels and hence conserve oil stocks.

4.3.114. Parker doubted that the hostility to nuclear power was capable of assessment. It varied in the population, and in strength from individual to individual; in some cases it could be dispelled by greater knowledge, in others not. Some fears would remain, 'no matter that those who feel them recognise them to be irrational'. He put down the anxieties as caused by reference in books and films, in particular films such as that of Wakstein's which was shown to him (84.85), to the Hiroshima bomb, and also by sometimes bland assurances of safety, such as MAFF's statements on the safety of discharges. He concluded, 'I am satisfied that although hostility does exist, it is not so widespread as to justify refusal of permission on that ground alone' (para. 13.1–2, 9). He set no store by opinion polls or petitions as he regarded the state of knowledge of the participants as not comparable to that of the local authority, and the questions in any case to be biased. In his section on public hostility he does not refer to the submissions of any of the national groups, such as BCC, nor to the evidence of FOE-WC, nor to the analysis presented by Taylor.

Case references
Bainbridge 60.55; Coates 64.90–98; Corkhill 60.7–16; Dalton 96.12 and 97.1–9; Dixon 60.40–52; Dudman 20.64–70, 24.44 and 67.88; FOE 93.7; FOE-WC 60.31–32, 39, 92–99 and 105–106; Goldsmith 85.33–62; Gosling 70.58; Hall 80.89–91; Halliday 70.94; Haworth 60.1–7 and 61–92; Higham 71.1–8; IOM 71.31 and 48; Little 47.79–82; Norman 60.18–22; PERG 96.4–29; Postlethwaite 70.79–93; SEI 86.43; Spearing 88.16; Spencer 86.20; Taylor 90.54–55; Wakstein 84.85

Proofs
Gosling; Postlethwaite; Taylor

Documents
FOE 134; PERG 32, 33, 35, 39, 40; TUC 2, 3

4.4. STATEMENTS OF INTENT

Waste Disposal and Discharges to the Environment

4.4.1. BNFL provided a number of statements of intent with regard to the environmental impact of their proposed activities (Proofs of Warner, Mummery, BNFL 195, 199, 230, 241, 300, 301, 302). Although technically speaking the Inquiry was to consider the THORP proposal, a great deal of reference to the Magnox plant was necessary, firstly because the impact of THORP would be in addition to that from future Magnox activities, secondly

---

Author's note: One hundred and sixty-four written representations were received (Parker Report Vol. 2, Annex 2, pp. 353–358.) No reference to these was made by Parker, and time has precluded an analysis here. Of these representations some were from County and District Councils, regional and national organisations, but the majority were individual objections.

because BNFL's intentions were based upon past experience with the Magnox plant (Proof of Shortis on design philosophy, and Warner on the evolution of a conceptual flowsheet).

4.4.2. As noted previously the Magnox plant had already been given planning permission for refurbishing. This refurbishing had been necessary in order to improve the decanning and increase throughput, and to further control effluent to the Irish Sea. In addition a Pond Water Treatment Plant (PWTP) would be built to reduce the caesium discharges from the Magnox ponds.

4.4.3. In this section we present details of BNFL's stated intentions with regard to discharge levels and to long term disposal of wastes. In the following section we look at the intended impact in terms of radiation exposure of the public and the workforce.

4.4.4. A number of tables were produced with discharge figures, and some were revised during the Inquiry (Mummery table 3, and revised table; Warner tables 3 and 4 with errata and alterations; BNFL 295). We reproduce the substitute table to Mummery for flowsheet discharges including margins in table 20.

4.4.5. Details of how the THORP discharges would be achieved are given by Warner (proof paras. 6–10, 18–21 for pond-water treatment, and 23–26 for expected effluents; 112–157 for process effluents, and 161–166 for potential treatment of gaseous isotopes; see also BNFL 86, 98, 126, 137, 138, 156).

4.4.6. In addition Warner (Appendix 2, paras. 98–101) provides details of potential flowsheet changes in the management of the MA effluent from the Magnox plant. In general HA and LA streams will receive the same treatment, but a number of options may be followed for the refurbished Magnox and for THORP; Warner emphasised that the designs were not fixed, that the flowsheet was at a conceptual stage. We present in table 21 a simplified version of Warner's table 1 from Appendix 2 as it illustrates the present Magnox effluent management and possible future revisions.

4.4.7. There was no clear statement of intent with regard to discharge figures themselves for both Magnox and THORP combined. Intent was expressed as targets for the percentages of ICRP dose limitations (cf. 4.5.123–129). However, in the confusion over expectations of reduced discharges, Mummery stated in answer to Wynne that the aggregate discharge figures of THORP plus refurbished Magnox would not be less than the current Magnox figures; in simple terms, the overall discharge in curies would not go down (20.50B). This matter was further taken up by Urquhart questioning Warner, who broadly agreed but was evasive as to the reasons, with the implication that 'other activities' would affect the future discharges (29.39GH–40E).

Table 20. Flowsheet estimates of radioactivity from the oxide fuel plant.

Radionuclide	Flowsheet estimated discharge including margins curies/annum
Liquid Discharges	
Strontium -90	1600
Zirconium -95/ Niobium -95	1400
Ruthenium -106	1600
Caesium -137 and caesium -134	8000
Alpha activity	1800
Tritium	1000000
Iodine -129	36
Aerial Discharges	
Strontium -90	1.0
Ruthenium -106	4.0
Iodine -129	0.064
Tritium	50000
Krypton -85	14000000
Carbon -14	600
Alpha activity	0.1

4.4.8. It was thus apparent that BNFL intended to reduce those discharges currently leading to fractions of the ICRP limit in excess of 10% (in particular caesium from the fuel storage ponds), but that due to the increased throughput of higher burn-up fuel, and possibly the military activities in addition, the total amount of radioactivity discharged would not be reduced.

4.4.9. With regard to discharges to atmosphere, the THORP plant would install the latest emission control devices which had proved their performance on the modern HAW tanks (Warner proof, paras. 157–166). For example the strontium retention factor was expected to reach a DF of 10^{10} or 10^9, and that for iodine-129 was estimated at 1000.

4.4.10. It was not envisaged that either carbon-14, krypton-85 or tritium gases would be retained. The possibility of retaining Kr-85 was being studied, but the Company had been advised that this was not necessary under the terms of the authorisation.

Magnox fuel separation plant, B.205/268. Basis: 5 teU/day, 3500 MWD/te, 350 day cooled fuel.

Effluent Stream	Volume m³/day	Curies per day arising					Disposal Route	
		Total Beta	Ru-106	Zr-95	Nb-95	Alpha	Present	Future
High Active Liquid Wastes	24	1×10^6	1.1×10^5	1.8×10^4	3.8×10^4	2×10^3	To HA liquor Evaporation and Storage (B215)	To HA Liquor Evaporation and Storage. To Waste Solidification.
Medium Active Acid, Salt-free Effluents	205	104	36	59	106	0.18	Via MA.3 Evaporator to delay storage (1.8 yrs) in B211 tanks. Then to sea via neutralisation plant.	Via MA.1 Evaporator to delay storage (1¼ yrs) in B211 tanks. Then to floc-precipitation and/or Ion Exchange for removal of Ce, Sr and alpha activities.
Medium Active Salt-Containing Effluents	39	130	64	42	76	0.14	Via MA Effluent Monitoring tanks and B242 to sea.	Via B211 Tanks to Salt Evaporator, B303. Concentrate to delay storage in B212 Tanks (5 yrs.) Then to floc-precipitation and/or Ion Exchange for removal of Ce, Sr and alpha.
Low Active Effluents	77	2.5	1.9	0.14	0.24	0.21	Via LA Effluent monitoring tank and neutralisation at B242 then to sea.	Via LA Effluent Monitoring tank and B242 to sea (some may go to Salt Evaporator)
Very Low Active Effluents	~ 300	< 5	4	0.5	0.5	< 0.1	To neutralisation at B242 then to sea.	To neutralisation at B242 then to sea.

Solid Waste Disposal

4.4.11. Clelland outlined the arisings of solid wastes that would require eventual disposal. BNFL's intentions were: (1) To solidify the HAW presently stored in tanks by use of a glass matrix. The process, known as Harvest (Corbet proof, paras. 11−19) had been developed on a laboratory scale and BNFL had been given planning permission for an industrial scale plant. The resultant glass blocks would require interim cooling, perhaps for several decades, in much the same way as HAW liquids, but would when cool enough, be disposed to geological formations: either the ocean or continental rock such as granite. The rate of arisings were expected to be 480 m^3/a containing 3000 MCi/a for glassifications, reducing to 150 m^3, 300 MCi/a for disposal (Clelland proof figure 1). (2) To solidify the MA sludges arising at a rate of 160 m^3/a containing 1 MCi/a, either with cement or bitumenisation, producing 200 m^3/a for disposal containing 0.01 MCi/a (Clelland proof figure 4). (3) Fuel element cladding (oxide fuel) would be treated with nitric acid and ultrasonics to remove adhering fission products, compacted and packaged for disposal. Arisings would be 600 m^3/a (25 MCi/a), which would reduce to 300 m^3/a (3 MCi/a) for disposal (Clelland proof figure 2). (4) Graphite sleeves from AGR fuel would be incinerated to reduce the volume and then packaged: arisings were expected to be 200 te/a (10,000 Ci/a reducing to 2000 Ci/a for disposal). No volume figures were given (Clelland proof figure 3). (5) In addition there are high alpha, low beta/gamma wastes currently stored in drums in special buildings. Arisings are 30 m^3/a (300 Ci/a), and it is intended to sort, incinerate, recover the plutonium by acid digestion, convert the ashes to ceramic blocks and package for disposal (Clelland proof figure 6).

4.4.12. All the above packaged waste would be in a form whereby it could be returned to overseas customers for eventual disposal.

Radiation Exposure of the Public and the Workforce

4.4.13. With regard to future doses to the workforce, Mummery states that it is the intention on THORP and Magnox to keep doses below 1 rem/a as an average for the whole workforce. The collective dose for THORP per year of reprocessing was intended to be 800 man rem (BNFL 302).

4.4.14. With regard to the environmental impact of aerial and marine discharges a number of documents were produced (BNFL 199, 295, 300, 301, 302, G 57 and CCC 16). These were summarised by Parker in terms of percentage of ICRP annual dose. We reproduce this as table 22.

4.4.15. For collective doses Parker summarises only those for THORP (Parker Report Annex 3). We therefore reproduce G 57 and BNFL 241 as tables 23−26 which give figures for THORP, present Magnox and future refurbished Magnox. Occupational doses of 800 man rem per year of reprocess-

Table 22. Environmental impact of discharges from Windscale expressed as percentage of present maximum permitted radiation doses.

Radio-nuclide	Critical pathway	1975 (Magnox plant)	Intentions for new plants		
			Refurbished magnox	THORP excluding (including) margins	Refurbished magnox plus THORP excluding (including) margins
Aqueous discharges					
Cs134/137	Fish	24	2.75	0.22 (1.1)	3.0 (3.9)
Sr90	Fish	2.0	1.1	0.05 (0.25)	1.2 (1.4)
H3	Fish	0.004	0.005	0.1 (0.1)	0.1 (0.1)
I129	Fish	0.04	0.06	0.5 (0.5)	0.6 (0.6)
	Total			0.87 (1.95)	4.9 (6.0)
Ru106	Silt	1.9	0.78	0.14 (0.7)	0.9 (1.5)
Zr95/Nb95	Silt	1.5	0.11	0.05 (0.25)	0.2 (0.4)
	Total			0.19 (0.95)	1.1 (1.9)
Alpha	Resuspension from silt (Inhalation)	0.13	0.04	0.02 (0.1)	0.06 (0.1)
Atmospheric discharges					
Kr	Immersion (skin)	0.08	0.11	0.9 (0.9)	1.0 (1.0)
H3	Inhalation and food	0.04	0.06	0.04 (0.2)	0.1 (0.3)
C14	Inhalation and food	0.06	0.09	0.15 (0.15)	0.2 (0.2)
I129	Milk	0.2	0.8	0.15 (0.6)	1.0 (1.4)
Ru106	Inhalation	0.14	0.14	0.05 (0.5)	0.9 (0.6)
Sr90	Milk	1.4	0.58	0.36 (3.6)	0.9 (4.2)
Alpha emitters	Inhalation	2.0	0.2	0.25 (2.5)	0.5 (2.7)
	Total			1.9 (8.45)	3.9 (10.4)
Totals all pathways		33.49	6.825	2.98 (11.45)	9.96 (18.4)

Notes:
1. The figures for refurbished magnox are based on (i) a throughput of 1500 tonnes uranium per annum, which is higher than ever before and the highest at which the plant is likely to operate, (ii) a cooling period of one year and (iii) the longest achievable time in the reactor—3500 megawatt days (MWD) per tonne uranium. The figures therefore cover the worst possible situation.
2. The figures for THORP are based on (i) a throughput of 1200 tonnes uranium per annum, i.e. the maximum theoretical capacity, (ii) LWR fuel one year cooled—an abnormally short period, (iii) the longest achievable time in the reactor—37,000 MWD per tonne uranium. Again, therefore, the figures cover the worst possible situation.
3. The 1975 figure for H-3 (tritium) is an assessment based on the measured discharge of krypton 85.
4. The 1975 figure for carbon-14 is calculated.

ing were given for THORP (BNFL 302).

4.4.16. Population dose commitments for aerial releases of strontium-90 and plutonium, iodine-129 and carbon-14, were also given (BNFL 300, G 57) and we reproduce these as tables 27 and 28.

Table 23. Collective doses due to discharges (with margins) and occupational exposures from THORP (PR Annex 3).

Nuclide	Document reference	Collective dose per year of reprocessing (man-rem)	Tissues or organ irradiated	Equivalent whole-body dose (man-rem)
Aqueous discharges				
Cs-134/137	BNFL 241	2141	whole-body	2141
Sr-90	BNFL 300	1600	bone	64
Pu (soluble)	”	1	bone	0
Atmospheric discharges				
C-14	G 57	5400	whole-body	5400
H-3 (tritium)	G 57	450†	whole-body	450
I-129	G 57	2800†	thyroid	112
S-90	BNFL 301	300	bone	12
Pu (if soluble)	BNFL 301	2650	bone	106
[if insoluble	BNFL 301	76	lung	20]
Kr-85	G 57	1750	whole-body	1750
Kr-85	G 57	370000	skin*	11100
Occupational	BNFL 302	800	whole-body	800
			Total	21935

Collective dose commitments are given per year of reprocessing (assuming 10 years of operation, with integration of dose commitments over 100 years) as quoted in the reference documents cited. Equivalent whole-body doses are derived in respect of fatal cancer induction according to ICRP weighting factors for individual tissues (its publication 26, para 105), relative to a fatal cancer risk of 10^{-4} per rem for uniform whole-body radiation (pubn. 26, para 60) corresponding to the non-genetic component of the whole-body weighting factor.

*For skin, ICRP consider that the risk of fatal radiation induced cancer is 'much less' than for other tissues considered (pubn. 26, para 63) including bone and thyroid. Here however a weighting factor of 0.03 relative to fatal cancer from whole-body radiation is applied (e.g. as compared with 0.03/0.75 = 0.04 for bone and thyroid).

†Includes dose from aqueous discharge.

Table 24. Population dose commitments attributable to discharges of Cs-137 and 134 from the Windscale works to the sea. (It has been assumed in each of the three cases that the plant is operational for ten years.)

Plant	Local 5 km	Local 50 km	National	EEC including UK	Rest of the world
Existing Magnox					
Annual (10th year)	70	1230	20200	36000	5
Integrated over 10 years	580	10700	170000	300000	43
Integrated over 100 years	660	12400	260000	460000	260
Refurbished Magnox (1500 te/y)					
Annual (10th year)	8	140	2400	4200	0.6
Integrated over 10 years	68	1250	20000	35000	5
Integrated over 100 years	78	1450	30500	53500	30
THORP (1200 te/y) (with margins)					
Annual (10th year)	3	60	950	1700	0.2
Integrated over 10 years	27	500	8000	14000	2
Integrated over 100 years	31	580	12000	21400	12

Table 25. Collective dose from refurbished Magnox (G 57) due to aerial discharges.

Annual collective dose (in man rem) in the 10th year of operation

Nuclide	Organ	Within 5 km atmospheric discharges	Within 50 km atmospheric discharges	National 1st pass	National global	National total	World
Kr-85	Whole body	0.10	0.52	7.7	1.7	9.4	117
	Skin	10	85	1500	360	1900	26000
H-3	Whole body	0.16	2.6	64	0.14	64	96
C-14	Whole body	0.17	5.6	150	30	180	2100
I-129	Thyroid	1.8	14	220	1.5	220	430

Collective doses (in man rem) received during the 10 year operating period

Nuclide	Organ	Within 5 km atmospheric discharges	Within 50 km atmospheric discharges	National 1st pass	National global	National total	World
Kr-85	Whole body	1.0	5.2	77	11	88	820
	Skin	100	850	15000	2400	18000	180000
H-3	Whole body	1.6	26	640	0.94	640	930
C-14	Whole body	1.7	56	1500	230	1700	16000
I-129	Thyroid	18	140	2000	11	2000	3600

Truncated collective dose commitments (in man rem) assuming constant discharge for 10 years (truncated at 100 years after the start of the operating period).

Nuclide	Organ	Within 5 km atmospheric discharges	Within 50 km atmospheric discharges	National 1st pass	National global	National total	World
Kr-85	Whole body	1.0	5.3	77	31	110	2100
	Skin	100	870	15000	6800	22000	450000
H-3	Whole body	1.6	26	640	1.6	640	970
C-14	Whole body	1.7	57	1500	460	2000	31000
I-129	Thyroid	18	140	2200	22	2200	4700

Table 26. Collective doses due to present Magnox (G 57) from aerial discharges.

Annual collective dose (in man rem) in the 10th year of operation

Nuclide	Organ	Within 5 km atmospheric discharge	Within 50 km atmospheric discharge	National			World
				1st pass	global	total	
Kr-85	Whole body	0.068	0.38	5.4	1.2	6.6	83
	Skin	7.4	60	1100	250	1300	18000
H-3	Whole body	0.12	1.9	47	0.10	47	70
C-14	Whole body	0.12	4.0	110	22	130	1500
I-129	Thyroid	0.46	3.5	102	1.1	103	250

Collective doses (in man rem) received during the 10 year operating period

Nuclide	Organ	Within 5 km atmospheric discharges	Within 50 km atmospheric discharges	National			World
				1st pass	global	total	
Kr-85	Whole body	0.68	3.7	54	7.8	62	580
	Skin	74	600	11000	1700	13000	120000
H-3	Whole body	1.2	19	470	0.68	470	680
C 14	Whole body	1.2	40	1100	160	1300	12000
I-129	Thyroid	4.6	35	875	7.5	882	2000

Truncated collective dose commitments (in man rem)
assuming constant discharges for 10 years
(truncated at 100 years after the start of the operating period)

Nuclide	Organ	Within 5 km atmospheric discharges	Within 50 km atmospheric discharges	National			World
				1st pass	global	total	
Kr-85	Whole body	0.68	3.7	54	22	76	1500
	Skin	74	610	11000	4800	16000	320000
H-3	Whole body	1.2	19	470	1.2	470	710
C-14	Whole body	1.2	41	1100	330	1400	23000
I-129	Thyroid	4.6	35	1020	16	1040	2800

Table 27. Population dose commitments in man rems for Sr-90 and plutonium, for ten years of operation and truncated at 100 years.

Population group	Existing Magnox (based on 1975 discharges)		Refurbished Magnox 1500 te/y 3500 MWD/te 1 year cooled		THORP (with margins)* 1200 te/y 37000 MWD/te 1 year cooled	
	strontium-90	plutonium	strontium-90	plutonium	strontium-90	plutonium
Local (0–5 km)	160	0.013	90	0.004	20	0.01
Local (0–50 km)	3000	0.27	1700	0.08	380	0.2
National	73000	7.7	41000	2.4	9200	6
EEC (including UK)	130000	14	70000	4.4	16000	11

*The margins on estimated discharges as quoted in Substitute Table 4 of Warner's proof have been incorporated.

Table 28. Truncated collective dose commitments (in man rem) for I-129 and C-14 assuming constant discharges for ten years (truncated at 500 years after the start of the operating period).

Nuclide	Organ	Within 5 km atmospheric discharges	Within 50 km atmospheric discharges	National			World
				1st Pass	global	total	
Present Magnox							
C-14	Whole body	1.2	42	1100	550	1700	37000
I-129	Thyroid	4.6	35	1020	22	2040	3200
Refurbished Magnox							
C-14	Whole body	1.7	59	1500	800	2300	51000
I-129	Thyroid	18	140	2200	31	2200	5300
Thorp							
C-14	Whole body	2.9	104	2700	1300	4000	88000
I-129	Thyroid	14	100	9000	240	9200	33000

4.5. INSTITUTIONAL SAFEGUARDS

4.5.1. In this section we examine the complex arguments relating to the controlling bodies. Several separate threads intertwine and considerable overlap existed between parties. However, the following threads have been followed here. (1) Whether the authorities were sufficiently open to a revision of basic radiation standards in the light of more recent knowledge; in particular whether the independence of the authorities can be assured. (2) Whether the authorities had been, or were likely to be sufficiently competent in assessing the implications of past or future activities. (3) Whether the system of control was sufficiently accountable and therefore acceptable to the public. (4) Whether in the light of all these points and of the history of BNFL's activities they were likely to meet their intentions with regard to environmental impact.

4.5.2. To some degree these distinctions are also arbitrary; for example competence and independence are inter-related, and whether BNFL meet their intentions may well depend upon future revisions of standards, past misinterpretations of the future behaviour of radionuclides already discharged and other such factors, as much as on whether the THORP control technology works as planned.

THE AUTHORISING BODIES AND RADIATION STANDARDS

Issue: Whether the authorising bodies were sufficiently open to a revision of standards.

The Health of the Workforce

4.5.3. BNFL and CCC as indicated above maintained that the ICRP was the right body in advising on standards, and that the NRPB was available for further advice should matters arise. As the chief matters of recent concern had been the Mancuso-Stewart-Kneale work, they had turned to the NRPB for advice on this matter. Schofield, for example, was advised that the amended Stewart table (TCPA 4B and IOM 66 table 11) was not significant (22.7—13). He stated however that he had also consulted the Company's own statisticians and that outside advice was also available.

4.5.4. With regard to the survey of Windscale's own workforce for excess cancers, the survey was carried out by the NRPB, and is reported in R54 (BNFL 119). Criticisms of this work are dealt with under questions of competence below (4.5.49—4.5.60). In particular a controversy developed over whether the R54 survey was adequate and whether a follow-up was worthwhile on epidemiological grounds. There had been disagreement within the NRPB, leading to the resignation of Goss, who felt that the findings were significant enough to justify further work (JKS 19). The Government had set up an Advisory Committee on the matter (on a National Registry of Radiation Workers), and the NRPB were represented on the committee by Pochin and McLean. Evidence produced by the TCPA showed that these two representatives held the view that a follow-up would have little epidemiological value, whereas other experts on the committee held opposing views, in particular Lindop and a member since deceased. McLean had given evidence to suggest that his view was representative and Lindop wrote to the Inquiry to correct this impression (TCPA 95, G 58, 49.13H–15, 74.28–30).

4.5.5   McLean, in answer to questions put by PERG, described how his view had also come to be that of the Government's as expressed in the response to the RCEP (BNFL 170, para. 9, Annex A). The White Paper had stated that a follow-up would be 'very costly and have little epidemiological value'. McLean stated that he had been telephoned by a department of the DHSS on the matter, and his comments over the phone had been repeated almost verbatim in the White Paper (49.13H). Taylor asked for details of the cost estimates, but none were produced.

4.5.6.   Taylor maintained that the NRPB were bounded by a philosophy of control that was outdated and a consequence of the fact that their management were ex-Windscale or Atomic Energy Authority safety staff (96.13). Indeed both Dunster of the Health and Safety Executive and McLean of NRPB had been UK representatives on ICRP (see Radford's critique below). Taylor was at pains to point out that his criticism was not on a question of integrity but one of a certain philosophy of control. This question was also raised by Atherley who stated: 'the impression overall from ICRP recommendations is that they are intended to safeguard the nuclear industry against extreme inconvenience of excessive safety precautions or fresh major expenditure on nuclear safety' (84.72CD). He felt that they did not recognise the spectrum of scientific views. Furthermore he held that the NRPB adoption of the ICRP economic criteria in standard recommendations was wholly wrong and that expert groups did not have the right to expound social judgements (84.64, WA 39, 81, 82, 84 and BNFL 114).

4.5.7.   Taylor quoted the RCEP's criticism of NRPB, and in particular the recommendation that the body be reconstituted at board level if it were to assume the role of statutory advisory body (see below). Radford was questioned on his critique of ICRP; the RCEP's conclusions were put to him, in particular that ICRP was a suitable source of recommendations provided its independence was assured. Radford, who held that ICRP was a closed group of experts who were continually ignoring new work and were not open to a revision of standards, remarked that such 'was a very revealing proviso' (75.81).

4.5.8.   Parker stated, 'I reject completely suggestions made that the control institutions were serving the interests of the nuclear industry in disregard of the public and the workforce'. He further held that such 'attacks' did nothing to further the case of those who made them and 'at times reached a level of absurdity which was positively harmful to such cases' (para. 10.130). He found that experience of the industry by those in authorities exercising control 'to be a clear advantage' (para. 10.131).

The Health of the Public

4.5.9. The applicant's position with regard to standards is set out by Mummery in proof (para. 7—28). He notes that with regard to exposure of the public the limit of 0.5 rem/a is deviated from by West Germany, where the limit for nuclear installations is 0.06 rem/a. Spearing provided evidence that the US standards were 0.025 rem/a for all sources of exposure (JKS 1, 2). As stated previously the Company maintained that their guideline was ICRP and that their intent was to remain at less than 10% in line with what was reasonably achievable. BNFL regarded this as an equivalent to the US approach.

4.5.10. There were a number of different nodal points at which parties submitted that basic standards were in question at creditable scientific levels, but that ICRP/NRPB were not paying sufficient heed. We have isolated the main arguments concerning, (i) the standard with regard to the genetic risk to populations subject to high collective doses but low individual dose, e.g. the Cumbrian, the UK and the EEC fish consumers subject to Cs-137 pollution; (ii) the question of the somatic risk from chronic low doses, particularly alpha-emitters; (iii) the question of the biological behaviour (transfer coefficients etc.), for radionuclides in the body or in the environment; (iv) the question of social judgements by expert groups on acceptable exposure standards.

Genetic Risk

4.5.11. In order to calculate the genetic risk to the local or UK populations two factors are important: a knowledge of the size of the population at risk (together with the total man rem collective dose to that population) and a risk estimate of genetic damage per unit of rem. From figures derived from BNFL's statements of intent Pochin calculated the impact of THORP (to the UK population?) according to the risk estimates of 1972 BEIR, and 1972 UNSCEAR reports, and 1977 UNSCEAR; the former two giving figures of 3 per 10,000 genetically significant man rems and the latter 2. Thus for THORP in each year of operation, between 0.7 and 1.0 substantial genetic defects would be caused in the total of all subsequent generations. For THORP plus refurbished Magnox and for the local population only, Pochin estimates a total of 229 genetically significant man rems per year of reprocessing, less than a tenth the previous figure, and leading to an average gonad dose per person per generation of 23 mrems, or one quarter of the UK limit of 0.1 rem indicated in the White Paper (BNFL 83). Calculations of past Magnox activity, or THORP plus Magnox exposures to the UK and EEC were not calculated (Parker Report Annex 4, 5).

4.5.12. Ellis (77.70) estimated a figure of 5.5 mrem per person dose commitment assuming the local population to be 100,000 (cf. Pochin 300,000) producing a figure of 0.360 rem per person per generation compared to Pochin's 0.023 figure.

4.5.13. Potts for the Lancashire and Western Sea Fisheries Joint Committee (LWSFJC), put a figure of between 30 and 50 severe abnormalities according to BEIR estimates as a result of past Windscale caesium discharges .(65.39, 55F).

4.5.14. Ichikawa provided a substantial proof on his researches on environmental radiation, particularly the effect on plants. He maintained that a reading of recent literature produced estimates of 15–20 rad as the DD for severe genetic abnormality, but that little was known about the hidden mutations and the effect on such subtle matters as overall 'genetic quality' (75.25–29). He also illustrated the effect of the UK standard of 33 mrem/person/annum if reached for the population as a whole; there would be: 170–840 fatal cancers; 10–100 serious genetic defects in the first generation; 62–1500 serious genetic defects at equilibrium; a 10% increase in the spontaneous mutation rate.

4.5.15. Spearing added that he thought the DD for hidden and polygenic mutations might be as low as 2 rads (87.7B). More generally, he felt that little was known of the effects of chronic irradiation at low doses and his proof contained the thesis that background radiation could not be viewed as negligible, there being evidence to suggest that some abnormalities, particularly Down's Syndrome, or mongolism, was prevalent in areas of high background (86.86–93, 87.1–41). The contention with regard to Down's Syndrome was advanced also by Urquhart who pointed to its prevalence in the granitic regions (higher background radiation) of Scotland and the Shetlands, where the rate was 1 in 714 as opposed to a national average of 1 in 2000 (89.22–23, 47, 75–76, JKS 6 p. 61).

4.5.16. In the light of these various estimates and doubts, Spearing took the view that ICRP underestimated the risk because of institutional problems: i.e. that geneticists were under-represented. Parker disagreed (para. 10.58–59).

4.5.17. A series of further views about standards were expressed, which for brevity we simply note as references below: Fremlin, Urquhart with regard to carbon-14, Radford with regard to tritium.

Case references
Dolphin 50.57–58; Ellis 77.70; Fremlin 39.13; Ichikawa 75.15–19; Potts 65.39 and 55F; Radford 75.90D–F; Spearing 86.86–93 and 87.1–41; Ur-

quhart 89.22—23 and 75—76

Proofs
Ichikawa; Spearing

Documents
BNFL 83; JKS 6

Somatic Risk

4.5.18. Risk estimates of the likelihood of developing a malignancy due to a unit dose of radiation (usually expressed as a frequency per million persons per rem with an expression time of 20 or 30 years) were recognised by all authorities as difficult to establish. As noted previously, early data were derived from populations irradiated by single high doses and it had been assumed that these dose–response curves (cancer induction per unit of radiation), which were linear at doses between 100 and 600 rads, could be extrapolated to very low doses. This hypothesis was for some time regarded as 'prudently pessimistic' but appropriate for the safety of workers exposed to doses of between 1 and 5 rems, and for the calculation of costs and benefits of discharge control technology. ICRP 22 (BNFL 149) recommended the adoption of cost-benefit optimisation based on this principle. Both BNFL and NRPB reiterated the point that the linearity of dose-response at low doses was a pessimistic hypothesis, and the ICRP Report 26, available at the time of the Inquiry for the first time, also reiterated this point (G 35).

4.5.19. This approach was criticised by a number of parties as not sufficiently taking into account recent work that confirmed the linear hypothesis (cf. Stewart 4.3.47—70), but it was also advanced that particularly with regard to alpha emitters (and possibly with regard to all chronic low dose radiation), the linear dose response might underestimate the risk. This latter view, exemplified by Morgan, a former Chairman of ICRP and BEIR committees, and member of USEPA working groups (NNC 14, WIERC 23), was put forward by Wynne, Urquhart and Spearing (21.77—79, 39.6—8, 25.92—93). Morgan had stated that in his view there was 'strong theoretical and experimental evidence that the linear hypothesis underestimated the risk from alpha particles'. Silsoe replied that the RCEP had considered the matter and their consultants had advised them that Morgan's view could not be substantiated (BNFL 9, para. 72, G 35, para. 30, and 76.12). In answer to Urquhart, Hermiston stated that the consequences of Morgan's view—a 200 fold reduction in the maximum permissible body burden (MpBB) for plutonium, would, if accepted, leave BNFL with great problems.

4.5.20. Apart from the general question of linearity, there was also a range of scientific viewpoints on the sensitivity of body organs to radiation doses. The ICRP gave a figure of 100 fatal cancers of all types per million persons per rem, and this figure has been used by BNFL for all calculations of consequences of exposures to workforce and public. However, Ellis pointed out that estimates varied and gave a table (Ellis, proof, table 1) with an upper figure of 120. He thought this might be extended to 150 or 200 by more recent epidemiological studies. Radford extended this to 170 on the basis of BEIR estimates and a 30 year expression time (as opposed to 20) to allow for very late cancers (proof para. 4.2, 75.80–81, NB this part not read into Transcript). However, on the basis of his recent studies, Radford maintained that the risk from chronic low doses was between 10 (for men) and 20 (for women) cases of malignancy per million per rem, the risk being that incurred per year after the time for expression had been allowed. He stated in clarification that this figure could be multiplied by the number of years at risk to derive the total risk per rem per million. This point was not entirely clear, but for lung cancer, Radford's estimates of 4 per million per rem per year could, he stated, be multiplied by 30, to give 120 (75.83D). (If the figure for all organs were so multiplied then the total risk for men would be 300 and for women 600, but it is not clear from the transcript that this is meant [Authors].)

4.5.21. Further documentation on risk estimates was provided by Urquhart: a paper by Thorne and Vennart of the MRC, whose uppermost risk estimate was 205 (WIERC 15).

Case references
Fremlin 39.6–8; Hermiston 25.92–92; Radford 75.80–83D and 76.12; Schofield 21.77–79

Proofs
Ellis; Radford

Documents
BNFL 9, 149; G 35; WIERC 15

Models of the Biological Behaviour of Radionuclides

4.5.22. In this section we are primarily concerned with those arguments relating to the revision of basic standards, rather than on questions of competence, and although there is here a degree of overlap, we deal with the main critique of environmental models in a later section (cf. 4.5.77–84).

4.5.23. Here we deal with one example in detail and draw passing attention to others. Radford in a detailed proof held that recent evidence put in question the ICRP/NRPB model of the behaviour of insoluble plutonium in the lung. His studies showed, he argued, that insoluble particles irradiated the sensitive bronchial epithelium to a higher degree than hitherto suspected, because of a non-uniform distribution of dose. There were several recent studies that corroborated his risk estimates, some unpublished (data on Czech miners, Swedish miners and ankylosing spondylitics), but a recently published paper also corroborated his work—that of Beebe in the US. Radford's current estimate was that the present ICRP dose limits for lung of 1.5 rem/a would produce 100% excess lung cancers (75.92G—H).

4.5.24. Radford had stated in proof that in his view the ICRP were 'reluctant to evaluate new evidence' (75.77), that they were a self-perpetuating group, drawn from radiologists with interests which affected their overall philosophy and recruitment to the committees assessing new evidence (74.83, 85—87). He also pointed out that Dunster, McLean and Pochin had served on such committees (see below). As an example, Wynne pointed out in re-examination that one of the papers on which Radford's view was based (the Beebe work) and which Parker had requested as supporting evidence, had been available to Pochin, who had not indicated the fact that it had been forwarded to him some time ago (76.5). Furthermore, according to Radford (75.92C), Dolphin had stated in evidence that there was no information to counter the view that the lung model used by ICRP was in error, in particular that the bronchial epithelium was at greater risk due to differential residence times (50.20A—21B). Radford felt this surprising as Dolphin had been Chairman of a session in Edinburgh when evidence was presented by Radford, and there had been data available at Harwell recently (75.92).

4.5.25. Radford indicated in his proof (75.77) why he felt the ICRP was prejudiced toward new knowledge. He outlined the process of scientific research and its evaluation, stating that all scientists have inbuilt bias that structures their research. He documented the gradual change in the approach of the ICRP (see below on questions of social judgements), arguing that it embodied a closed and not wholly independent world view. This structured its approach to new knowledge (75.77—81).

4.5.26. Taylor (96.13—15) and Wynne (94.63—70) both argued that the ICRP and the NRPB by virtue of their respective philosophies originating in the service of the industry, could not be viewed as bodies open to a revision of standards.

Case references
Dolphin 50.20A—21B; NNC 94.63—70; PERG 96.13—15; Radford 75.77—81, 83, 85—87 and 92, 76.5

Proofs
Radford

The Acceptability of Exposure Standards

4.5.27. In this section the following broad categories of argument are illustrated: (i) the implications of social judgements in the working of expert groups concerned with the recommendations of exposure limits, e.g. ICRP, NRPB and MRC; (ii) the presentation of these standards to the workforce and to the public in various forms; (iii) the interpretation of these standards with regard to radioactive discharges to the environment, i.e. plant waste management and discharge control strategies.

4.5.28. As stated above the ICRP standards incorporate a social judgement with regard to what would be an acceptable exposure to a workforce and to the public. Thus within the UK standard setting system, social values are inbuilt at the very beginning, and as pointed out by Radford, Taylor and Wynne, the bodies set up to advise on these standards also contain influential members of the committees that originated them, e.g. Dunster, Pochin, McLean. Various parties criticised this system. As we saw above, it was felt that it unduly prejudiced the standard setting bodies with regard to emerging studies that might seriously inconvenience the industry—either the radiologists themselves, or the power generation programme.

4.5.29. The major critique of this incorporation of social judgements was made by Radford (75.34—90, 76.5, 34, 66, 67) and by Atherley (84.64, 72C–D), and were part of the cases of NNC and PERG. Radford outlined how the recommendations of ICRP with regard to discharges and exposures had altered over the past decades to reflect changes in their social (and institutional judgements). Thus a four person committee, which included Dunster, had altered the rules from the original, which were to keep exposures to as 'low as practicable' with unecessary exposures avoided, to 'provide for operational control of specific procedures, ... resulting doses are as low as reasonably achievable, economic and social considerations taken into account'. Radford maintained that the direction of change was wrong in the light of public concern and the global implications of discharges. He argued that such a closed and self-perpetuating body should not make aconomic and social judgements, but should confine itself to the scientific task of producing risk estimates. Whether the risk from exposure (and there would need to be sufficient plurality of research institutions to ensure that the risk assessments were independent of vested interest) is acceptable or not, is for a political body to decide. Such a body would of necessity have an apparatus for the participation of those at risk (75.89).

4.5.30. This argument was also pursued by Atherley (cf. 4.5.62), and reiterated by Spearing and Ichikawa (75.23). Wynne for NNC applied the arguments to the setting of discharge controls and the system of authorisations in the UK, arguing for plurality in research and a separation of functions (94.63—66, 68—70, 75).

4.5.31. Further elaboration of the nature of the social judgements was contained in the assumption by ICRP/NRPB and followed at numerous points by BNFL and CCC, that exposures from wastes could be deemed acceptable if they were but a fraction of natural background and within the normal fluctuations to be found at various points in the UK. For example, the north-east of Scotland has almost twice the natural background experienced in parts of Southern England. Parker saw this as clearly relevant (para. 2.29). However, apart from the arguments documented above, that natural background could not be deemed to have a negligible effect, the issue at stake was whether an expert group such as ICRP should incorporate such judgements into standard setting.

4.5.32. In one area this trend was apparent. Laxen and Wynne, in cross-examination of Webb of NRPB and Dunster of HSE, elicited the view that a dose-cut off should operate where large populations received very small individual doses (G 34). Wynne indicated that a cut-off at 1% of background as was advocated would alter the cost-benefit calculations in gross favour of the industry and against the installation of expensive control technology for such radionuclides as krypton (NNC 16, 20 and 50.1−5, 45.79), and for caesium (20.51−57).

4.5.33. With regard to the presentation of standards, Radford's view is documented above: that such bodies as ICRP should serve a simple function of risk assessment and not 'set' exposure limits. However, there was some dispute as to the best way of presenting risk assessments. Radford put forward the following: that figures such as malignancies per million persons per rem were meaningless to the man in the street. He argued that risk should be expressed as the extra risk of cancer due to the exposure. Thus it might be expressed as a percentage increase. He personally felt that the percentage increases resulting from BNFL's activities, ranging from 8% for local female fish consumers, to perhaps higher figures for Ravenglass villagers, were not acceptable, and that the EPA standards which approached 1% were acceptable, involving as they did considerable participation in the process of weighing of costs and benefits. However, Radford admitted there was a problem when public knowledge of the background cancer rate was inadequate. Parker felt that Radford's view would present meaningless figures and preferred the comparative view where risks of this nature are compared to smoking, driving cars, climbing mountains etc., i.e. expressed as one chance in so-many-thousand of contracting a fatal illness. Parker states: 'if a person is told that, as has been estimated, such risk (1 in a million) is the same as that involved in smoking 1½ cigarettes, travelling 50 miles by car or 250 miles by air, rock climbing for 90 seconds, canoeing for 6 minutes, engaging in factory work for 1−2 weeks or simply being male aged 60 for 20 minutes, it would mean a great deal to him' (para. 10.29).

4.5.34. This view had been put forward by Fremlin and was presented by the RCEP (BNFL 9, para. 170–173). Fremlin compared the various Windscale discharges and consequent exposure risk to the smoking of cigarettes, and had advised the CCC (as consultant) on the acceptability of the risk in these terms. For example, he maintained that 1% of ICRP limit for the public was equivalent to smoking one cigarette per year. Another example of his comparative view was that Windscale fish containing radioactive caesium were, he felt, safer than tinned luncheon meat. He also presented the usual expression of extra chance of malignancies, maintaining that 30% ICRP was equivalent to an extra 1 in 60,000 (33.9–13).

4.5.35. Parker carries this argument further (para. 10.30–10.34). On the comparability of risk he states that he was supported by Wynne, in that there was no better way. However, he does not call on the evidence submitted to him by Taylor and Atherley. The former presented in proof the results of an IAEA study on comparability of risk which showed that the mode of dying was of paramount importance: that the components of attitudes to risk were complex, and that the public operated a hierarchy of values (Taylor proof, para. 89, 100–102, and PERG 33). The IAEA sponsored study showed that 'the risk methodology fails to consider conceptual differences in how risks are perceived by the public and by those compiling the statistics' (PERG 33, Taylor proof, para. 101).

4.5.36. However, Parker goes on to provide comparisons: a member of the public exposed to 500 mrem (the ICRP limit) will have a 1 in 20,000 risk of dying, and would be ten times more likely to die from an accident of some sort in the year of the dose rather than of cancer at some later date (para. 10.32). He concludes, 'neither risk appears to be severe'. Thus if BNFL maintain their intentions of limiting exposures to 10% of ICRP, an individual so exposed has a 100 times greater chance of dying of something else, and 'if he smokes 10 cigarettes a day, he will be 500 times more likely to die from that cause than from THORP and Magnox combined'. The annual risk from this source will be about the same as that involved in travelling 250 miles by car or being male aged 60 for 1 hour and 40 minutes' (para. 10.33). Parker makes no reference to the evidence on public perceptions of risk, and concludes, 'it seems to me impossible to suggest that any substantial numbers of the public or of workers would regard the risks as intolerable'.

4.5.37. Taylor presented detailed evidence that substantial numbers of the public would not eat fish if they knew it to be contaminated, in spite of reassurances from 'scientists' (PERG 35); he also presented evidence of attitudes to environmental contamination generally (proof 81–99). The *New Society* Survey (PERG 35) found that more than half the female population (slightly less for males) would not eat contaminated fish. Taylor also pointed out that the ICRP Report 22, when advocating the use of cost-benefit optimisation, had stated that there should be some process of allowing for aesthetic and other human factors when considering the detrimental impact of discharges on food supplies (47.38–40, 51.12, 96.9C–E). Taylor argued that the so-called 'irrational' behaviour of the public with regard to risks it was prepared to undergo and those it was not, was not something to be dismissed, and more fundamentally decisions taken on behalf of the public without adequate participation (see later sections), were likely to be rejected once the public became fully aware. This latter statement was supported by sociological evidence on the antinuclear movement in Europe.

4.5.38. In argument relating to these issues Urquhart compared the standards in use by BNFL to those in use for other pollutants such as mercury. He maintained that were BNFL to operate to these standards then their discharges would have as their maximum limit 1% of ICRP as exposure impact. This would mean a discharge for caesium of 720 Ci/a as limit, and a plant working level of 140 Ci/a. For plutonium the discharge would be limited to 0.004 Ci/a. He maintained (see below) that foreign reprocessing plants had substantially lower limits, especially for alpha: La Hague had a limit of 90 Ci/a and a plant discharge of 27 Ci/a (cf. Taylor and PERG 1 for projected German discharges).

4.5.39. With regard to the workforce, Wynne concluded for NNC (94.72), that he was 'not at all convinced that the workforce understood to an adequate degree the implications of their day-to-day exposures'. He contrasted the language of documents used by management in training programmes (NNC 4, BNFL 18), with the language of scientific documents (NNC 5). The former continually spoke in terms of 'safe' levels, giving the appearance of permanence, and not the true situation which was one of uncertain and shifting standards.

4.5.40. Mummery had replied to these questions (20.29–30): 'It is important that the people who talk to the staff have some discretion in the words they use . . . in order to get the message over.'

4.5.41. Glidewell for CCC had put this question to Fremlin who was of the opinion that the workforce did appreciate the risk 'in an adult way'. Fremlin referred to the transcript of a public meeting in Cumbria (CCC 12) where Maxwell, a Union representative from the plant, had stated that his colleagues were 'men of metal' and had rejoindered 'God save us from experts' when the question had been raised (39.46–48).

4.5.42. Wynne also raised questions as to the workforce's understanding of the genetic risk (66.100–101).

4.5.43. The system of standard setting was thus seen by the objectors to discharges as being the source of the rationale for discharge targets. It was also seen to affect attitudes to a number of matters that we deal with under questions of competence. Here we outline the form of the argument as it relates to the openness of the authorities to considerations of alternative philosophies of control.

4.5.44. It was a main contention in the case of NNC that the ICRP-dominated philosophy affected several key aspects of monitoring: for example, the change in the formulation of critical groups in the population at risk; the absence of validation studies of models for the transfer of radioactivity through the food chains to man; the research base for model formulations; and the response to situations where controls failed, such as caesium releases from corroded fuel elements and the build-up of alpha emitters at Ravenglass.

4.5.45. With regard to critical groups Thompson, Wynne and Laxen highlighted the way the formulation had altered in such a way that the most highly exposed individual was no longer the limiting factor, and that an averaging procedure had been introduced (67.44−45, 47−48, 19.41, 20.43−45, 26.92−94, 69.38, 42−57, 67−68, 70.44). Silsoe replied that provided the exposures were a few per cent of ICRP then the safety factor was such that the critical group was legitimate (9.17−18, 27.6−9). NNC's case was that as percentages had now risen to 35% and perhaps even 50% of ICRP in the critical groups according to the model, then validation studies should have been carried out. This prompted Parker to order a series of tests on local fish consumers, should they wish it (they had never been approached). We present the results under questions of competence below.

4.5.46. NNC also held that the response to caesium control (cf. 4.3.31−39) was belated and indicative of a laxness and lack of understanding of what was acceptable pollution. Both the LWJSFC and the IOM found the caesium levels in fish unacceptable, even though the ICRP limit had not been reached (93.35, 94.68−70, 75). Ellis also regarded the collective doses due to caesium as unacceptable (77.49, 50).

4.5.47. NNC were critical of the MAFF concept of Limiting Environmental Capacities (LEC) which related to ICRP standards. This had led to activities which, seen in the light of an assumed LEC, were perfectly reasonable, but that if a pathway emerged which led to higher concentrations and doses to man, then activities (discharges) might not be amenable to rapid alteration; in the case of alpha emitters in Ravenglass the wastes had already been dispersed and their future behaviour could not be easily affected. NNC showed that BNFL had been caught out with regard to LEC and the Porphyra pathway, the Ravenglass silt and the resuspension of plutonium (cf. 'Questions of Competence' below).

Case references
Adams 66.100−101; Atherley 84.64 and 72C−D; BNFL 9.17−18 and 27.6−9; Dunster 45.79; Ellis 49.50; Fremlin 33.9−13, 39.46−48; Hermiston 26.92−94; Ichikawa 75.23; IOM 93.35; Laxen 69.38−57 and 67−68, 70.44;

Mitchell 47.38−40; Mummary 19.41, 20.29−30, 43−45 and 51−57; NNC 94.63−66, 68−70, 72 and 75; Radford 75.34−90, 76.5, 34 and 66−67; Thompson 67.44−45 and 47−48; Webb 51.12

Proofs
Radford; Taylor; Wynne

Documents
BNFL 9, 18; CCC 12; G 34; NNC 4, 5, 16, 20; PERG 33, 35

THE COMPETENCE OF THE AUTHORISING BODIES

Issue: Whether the competence of the controlling bodies was sufficient to assure an adequate assessment of the risks and to create confidence that future intentions would be achieved.

4.5.48. The competence of the controlling bodies was brought into question on a great many issues and there are many strands to the arguments. We have identified the following categories as the main strands and have not attempted to reference every technical point, but only those that led to conclusions in the main argument of a party.

(i) As a great deal of the assessment of exposure risk to both the workforce and the public depended on a mixture of monitoring data and models of the behaviour of radionuclides either in the environment or the human body and, further, on assumptions as to the biological effects of low doses, a basic criticism put forward concerned the use of validation studies. In the case of the workforce, for example, this would mean epidemiological studies on cancer rates (or other health effects) as a source of direct information. This could also apply to studies of the surrounding population, although direct correlation to dose would be more difficult. In the latter case, it was discovered that local critical groups had never been monitored for their actual uptake of radionuclides. As will be noted, when this had been done elsewhere, models of uptake had been found to be in error. Parker instigated a number of *ad hoc* tests during the Inquiry involving the use of BNFL's Whole Body Monitor.

(ii) The strategies of monitoring in the environment were put under detailed scrutiny. It was found that the authority concerned, MAFF, due to limited resources, had to make strategic decisions on where to monitor and in how much detail. From such sampling, MAFF drew up a picture of the most limiting pathways back to man (those exposing the critical groups). There were therefore gaps in the monitoring data from year to year, for example of certain organisms acting as indicators of trends. According to the picture drawn up, the capacity of the environment to absorb a given discharge without exposing anyone to the ICRP limit was calculated; this was known as the Limiting Environmental Capacity (LEC). In order for this calculation to prove accurate, it could be seen that MAFF must be able to predict the future behaviour of radionuclides in the environment, including those with very long half-lives. These predictions must also include the possibilities of new pathways developing (as at Ravenglass for example), synergistic effects, and the possibility that basic biological data on toxicity and dosimetry might change. For monitoring purposes there must also be working limits and a system of feeding back the implications of sample levels to the plant management; Derived Working Levels (DWLs) were therefore used for certain key organisms in the food chain, and for silt, soil etc. (In this section we have also included the general 'housekeeping' on the site itself as this reflects the overall response to monitoring data and safety in general.)

(iii) In the case of discharge control technology, and waste storage or disposal strategies, the question of awareness of alternative procedures arose. The philosophy of the Company was reflected in the technology of its discharge control and future intentions on the disposal of wastes, and likewise the competence of the authorities was reflected in their overall awareness of the technological alternatives and in their will or ability to induce the Company to follow alternative strategies or to make provision should a chosen strategy fail (as for example the wet storage of Magnox fuel).

(iv) Questions were also raised as to the suitability of the site. This was an awkward issue as it was contended that the Inquiry should have been able to consider alternative sites and detailed environmental impact assessments for each, whereas no such evidence was available. The Inspector therefore implicitly accepted BNFL's contention that the infrastructural arrangements and advantages outweighed the disadvantages of the site with regard to the enclosed nature of the Irish Sea or the proximity of the National Park. BNFL admitted, however, that it had earmarked future sites should a second THORP be necessary (cf. 4.5.104–106, 6.2.8–15).

(v) The competence of the controlling bodies, it was argued, was in many ways a function of their organisation. The integration of basic research, monitoring results, and discharge control procedures was found to be dependent on a number of separate bodies, a matter criticised by the RCEP, and this was debated at some length.

Protection of the Workforce: Health Statistics

4.5.49. As noted previously BNFL followed the guidelines laid down by the ICRP as advised by the NRPB. Criticism of the controlling bodies thus centred upon these bodies. We are particularly concerned in this section with the question of validation (basic standards are dealt with above). As discussed below the RCEP had requested a survey of cancer statistics for the Windscale workforce and this had been carried out by the NRPB. Their report, NRPB R54 (BNFL 119) was relied upon in evidence by BNFL (Mummery, proof, para. 62, and 21.84–87, 99.58) and by CCC (97.73G–H, 98.6–9).

4.5.50. R54 was the only evidence of a statistical nature presented by BNFL or CCC on the correlation of cancers and exposure in the workforce. CCC did not cross-examine the NRPB witness and did not alter their position in their closing statement.

4.5.51. Several parties produced evidence and argument that this report showed incompetence on the part of the NRPB.

4.5.52. It was pointed out by Taylor in cross-examination of Schofield that R54 was the published version of the report submitted at the request of the RCEP for their sixth report (BNFL 9). The RCEP had found the NRPB study inadequate in that approximately 50% of the potential data on cancer deaths had not been included (BNFL 9, paras. 74–75, and 223–224). In spite of this R54 was published and distributed to the public as an 'interim report'.

4.5.53. Taylor also pointed out that when considering whether the NRPB should have a statutory function in advising the Government, the RCEP had stated that for it properly to do so, it would need to be reconstituted at Board level. They also called for a review of the organisation and expertise of the executive body (BNFL 9, para. 224).

4.5.54. Alesbury, with the aid of Blackith, a Fellow of the Institute of Statistics, showed that R54 contained methodological and arithmetical errors which, he maintained, led to a significant understatement of the correlation of cancers and exposure in the workforce (49.22–47, 50.11–15, 61.106–109, WA 27, 66, 69). Dolphin admitted the arithmetical errors and agreed that further work should be done, but in the case of the Myelomas affected by these errors, which now appeared to be significantly correlated with radiation exposure, he advanced the hypothesis that there was some other common cause as these persons did not appear to have been highly exposed (51.14, 49.29E). Alesbury argued that the RES neoplasms now appeared between five and ten times more likely than chance to have been caused by occupational exposure, he called it a 'strong finger of probability' (92.28–30). Pochin warned that the figures in question could be 'a freak of chance' (92.36–42).

4.5.55. Parker concluded, 'I have no doubt about the importance of such data' (i.e. the analysis of those workers who had left the industry). 'I say no more on the subject, except that I have not relied on the paper by Dr Dolphin [R54] reviewing figures relating to Windscale workers in reaching my conclusions on Dr Stewart's evidence, nor indeed on any other matter' (para. 10.47).

4.5.56. Stewart referred to R54 as containing 'grave defects' and 'inappropriate statistical tests' (74.22−27). In addition the errors, which were immediately obvious to statisticians, were not picked up before publication because the report was not sent out for peer review. Alesbury noted that S. Goss, a former NRPB statistician who had resigned his post, had pointed out some of the errors in a letter to the *Observer* newspaper in March 1977 (JKS 19), but that R54 had not been withdrawn or corrected and had been used in evidence. Goss also stated that the figures justified further investigation.

4.5.57. In response to Dolphin's criticism of her work (50.5−7) Stewart had replied that he clearly did not understand statistics (74.25). She also pointed out that her method had been adopted by the forthcoming official analysis of the Windscale data by Smith in a survey independent of the NRPB.

4.5.58. Dolphin was cross-examined by Taylor on the assumptions upon which R54 was based (50.24−27). Dolphin had assumed that: (i) exposure was constant over the period studied; (ii) workers left employment at a constant rate; (iii) the risk per year after exposure was constant and lasted for 20 years. He admitted that these assumptions were not based on any evidence, and that if one took alternative assumptions, e.g. that exposure was greater in the earlier years of the industry, that workers left at a non-constant rate, and the risk after exposure was not constant, then the significance of the lost data would be affected.

4.5.59. Evidence that the 'lost workers' might significantly alter the picture was provided by Stewart who pointed out that BNFL workers were allowed to convert pension rights into a lump sum. There would therefore be no incentive for ill workers to remain in the industry and this would lead to a bias within the group of pensioners. She stated, 'the clue to the deficiency of observed deaths in the Windscale survey probably lies in the exclusion of all ex-employees from the study' (74.28−29, and cf. 45.78).

4.5.60. As noted previously it was contended that the NRPB did not properly encompass the views of some experts that the data on the missing workers was worth retrieving for epidemiological reasons. Pochin believed that any effect of exposure on these workers would be masked by the background cancers to be expected in older people (74.30, 49.13H, 63.38−45, 92.32−36), and he had sat on the same Advisory Committee as Maclean (TCPA 95, G 58) at which other members had expressed opposing views. However, Maclean had presented his own view and that of Pochin to the Government and this had become incorporated in the Government's answer to the Royal Commission (cf. para 4.5.4−6). Schofield had also been advised in a similar vein (22.7−13).

Protection of the Workforce: Exposure Standards

4.5.61. Laxen and Urquhart charged that the NRPB when faced with a range of risk estimates consistently took the lower figures for the purpose of modelling effects (50.18−24, 49.15). For example, the Relative Biological Effectivity of plutonium was taken as 10 whereas some research workers would use 25 (WIERC 15). There was also a question of residence time in the lung in relation to smoking. Laxen also asked Dolphin whether their standards were based on those individuals or groups in society who might be radiation-sensitive; Dolphin answering that it was right to take the mean (50.24, NNC 43 para. 426). Radford later commented that such an attitude by a key member of the NRPB 'displays an appalling lack of understanding of current procedures in establishing environmental protection standards' (75.86F−87A).

4.5.62. Atherley criticised the NRPB's acceptance of the economic criteria in standard setting (BNFL 114, R46 by Pochin, Maclean and Richings), when in relation to Windscale the contracts were 'cost-plus' and therefore economic criteria were irrelevant and a 'physically possible' standard should be used (84.65). This attitude with regard to economic criteria was debated at greater length in relation to risks to the public from marine discharges, Wynne and Laxen arguing that the introduction of social judgements was inappropriate for an advisory body on standards and contrasted strongly with US procedure (cf. 4.5.30, 6.1.5−9).

4.5.63. Wynne commented on NRPB attitudes: 'NRPB tried every criticism to rebut Stewart−and failed. Would they had shown similar critical resolve toward documents proposing relaxed standards' (94.71).

4.5.64. Urquhart maintained that NRPB should be looking at the Ischaemic heart disease prevalent in the Windscale workforce (Mummery table 1, cf. 4.3.48−49). He referred to WIERC 56 which indicated a possible correlation of alpha-emitters and heart conditions. Dolphin replied that this was not one of NRPB's priorities (50.38).

4.5.65. Parker's overall conclusions are contained in a section on the quality and integrity of the advisory and control authorities. He stated that he regarded 'attacks' on the 'integrity' of these organisations as without foundation (para. 10.130). His observations on competence were limited in the case of the NRPB to this question of Dolphin's paper (paras. 10.133, 136). On Dolphin's admission of a mistake in the methodology, he concludes, 'I cannot regard this matter as indicating that NRPB are incompetent', basing his conclusion on the argument that few people never make mistakes, and that if one such mistake rendered a person or body incompetent, then were there complete disclosure of all facts, we would probably have to regard all experts and all bodies to which they belonged as incompetent (para. 10.136).

Protection of the Workforce: General Housekeeping

4.5.66. The RCEP had concluded that during the time of their visit in 1974 'housekeeping' was not of the highest standard, and had recommended that the then new management should give attention to this (BNFL 9, para. 368). Several parties referred to incidents at the plant.

4.5.67. Wynne when cross-examining Hermiston cited several examples of contaminated clothes and equipment finding their way off-site, in particular an AGR reactor plug was found on a non-active tip (26.85); there was an incident in which the car park was contaminated (26.86); and he questioned the discipline regarding the wearing of badges and the recording of doses in relation to incentive schemes and bonuses, derating of workers after exposure etc. (26.87–89). Hermiston replied that there was no 'bonus' scheme, but an 'irksome conditions allowance' (26.89).

4.5.68. Atherley, a witness for WA with experience in the training of Inspectors for the Health and Safety Executive, concluded: 'existing provisions at Windscale for control of risks to health and safety of the community at large and to employees are not sufficiently stringent'. His chief criticism related to the economic criteria used for the installation of safety equipment, shielding etc., and to the attitude of management and the authorising bodies defining 'acceptable risk' as opposed to arriving at a consensus involving extensive participation of the workforce (84.64–93, WA 39).

4.5.69. Ellis was critical of BNFL's record on exposure to the workforce: 'these higher levels are probably the highest levels of exposure reported in any industry involving radiation in the UK or elsewhere in the world' (77.45F). On internal contamination there was a 'high frequency, that I personally would not regard as an acceptable situation' (77.45G). Silsoe replied that certain other industries in the UK did in fact have comparable figures; that BNFL was already reducing the higher doses, and intending incorporating increased shielding in THORP as well as the new Magnox plant; and that other countries could perhaps not be relied upon to present such extensive data on exposures as did the UK (77.58–71).

4.5.70. Higham raised questions on the washing decontamination procedures and enforcement (23.4). Urquhart raised the problem of setting trigger levels for air-monitoring (25.55–56). Barratt questioned on the subsequent fate of workers over-exposed in accidents (19.34) and on various glove-box incidents of contaminated wounds (21.93–97, 101–104).

4.5.71. Dudman raised the question of the mental effects of working with radiation, in particular of those workers undergoing decontamination treatment in relation to their knowledge and understanding of the risks (21.66–67). Schofield replied that 'occasionally we get someone who really is scared to death and we wonder why they came here in the first place'; however, such 'neurotics' were not dismissed (21.89–92).

4.5.72. Adams, a national representative of the EETPU, stated that his union was satisfied with the safety arrangement and record of BNFL in protecting the workforce. He maintained that the nuclear industry was exemplary and that the Trade Union Movement could be relied upon to make sure safety was not sacrificed for production needs (66.90G–92C). Questions as to how representative this view was in relation to BNFL's strike record were raised by Collins for FOE-WC (66.93H-98A).

Case references
Adams 66.90E–92C; 93H–98A; Atherley 84.64–93; Blackith 61.106–109, 63.38–45; BNFL 99.58; CCC 97.73G–H and 98; Dolphin 49.27–47, 50.5–7, 11–15, 18–27 and 38, 51.14; Dunster 45.78; Ellis 77.45 and 58–71; Hermiston 23.4, 53–56, 26.5–6 and 87–89; Maclean 49.13H and 15; Mummery 19.34; NNC 94.71; Radford 75.86F–87A; Schofield 21.66–67, 84–87, 89–92, 93–97 and 101–104, 22.7–13; Stewart 74.22–30; WA 92.28–30, 32–42

Proofs
Blackith; Mummery; Schofield

Documents
BNFL 9, 114, 119; G 58; JKS 19; NNC 43; TCPA 95; WA 27, 39, 66, 69; WIERC 15, 56

Protection of the Public: Whole Body Monitoring

4.5.73. The criticisms relating to a lack of validation studies in those locally exposed groups where the dose had reached a significant fraction were made chiefly by WIERC, NNC and IOM. The Company's view was expressed by Mummery, who held that it would have been unreasonable to have subjected the public to tests. However, he admitted that they had never been offered the facility (19.43). Reasons for the importance of validation were exemplified by Urquhart, who provided evidence from Tarapur in India, where large discrepancies had been found between doses predicted by modelling, and those predicted from actual measurement of the body content of the population (26.37–42, 17–22, WIERC 25, 31).

4.5.74. Toward the end of the Inquiry, Parker offered WBM facilities to local witnesses who were worried about the potential hazard from consuming local fish. Arrangements were made for a test, agreed by Haworth of FOE-WC and advised by Fremlin of CCC. The results were presented as BNFL 326, as illustrated in table 29.

4.5.75. Haworth's comments on the figures are recorded during his closing speech when the document was first made available (94.14C–16).

Table 29. Whole body monitoring of volunteer local fish consumers.

Subject number	Reported average quantity of fish consumed per week (ounces)	Whole body Cs-137 (nCi ± 95% confidence limit)			Estimated equilibrium values*	
		Initial	2nd Week	4th Week	nCi	% ICRP Dose Limit
1	10½	0 ± 2	19 ± 2	22 ± 2	122 ± 13	4
2	10	9 ± 2	18 ± 2	21 ± 2	80 ± 13	3
3	12	6 ± 2	15 ± 2	23 ± 2	101 ± 13	3
4	10	10 ± 2	25 ± 5	25 ± 6	94 ± 36	3
5	9	2 ± 2	N/A	16 ± 2	84 ± 12	3
6	12	2 ± 2	29 ± 6	32 ± 6	173 ± 36	6
7	7½	24 ± 6	30 ± 6	34 ± 6	29 ± 6	1
8	14	9 ± 2	28 ± 5	33 ± 6	144 ± 36	5
9	14	14 ± 2	37 ± 6	37 ± 6	146 ± 37	5
10	12½	3 ± 2	21 ± 2	26 ± 2	135 ± 13	4
11	8½	13 ± 2	23 ± 2	21 ± 6	59 ± 32	2
12	6	35 ± 6	39 ± 6	38 ± 6	38 ± 6	1
13	6	42 ± 6	53 ± 6	51 ± 7	49 ± 7	2
14	33	246 ± 10	257 ± 9	227 ± 10	243 ± 10	8
15	6¾	29 ± 6	39 ± 6	37 ± 6	35 ± 6	1
16	6	10 ± 2	18 ± 2	16 ± 6	41 ± 34	1
17	9½	9 ± 2	27 ± 6	28 ± 6	116 ± 36	4

* Assuming 100 days effective half period–MRC.
1 N/A denotes not available.
2 Subjects number 7, 12, 13, 14 and 15 were considered to be at continuing equilibrium. For the subjects who were not considered to be at equilibrium the estimated equilibrium value as shown could be too high or too low by up to about 50% depending on the actual half period.

Although he states that there is need for further tests to ensure proper sampling and continuity, he found the levels gave no immediate grounds for concern. He added, however, that the very nature of radiation was such that considerable doubt and fear would remain in the local community as a result of the various expert disputes about radiation standards.

4.5.76. Parker concluded by quoting Haworth's statement of assurance and calculating the ICRP limits for various extensions of the data available, for example if the maximum consumer identified by MAFF (830 g/day) were to consume fish from these test runs, then he would reach 61% of the ICRP limit (para. 10.85–10.95).

Protection of the Public: Strategies of Environmental Monitoring

4.5.77. It was held by a number of parties that the authorities' knowledge of the environmental behaviour of radionuclides was not adequate for the purposes of the prediction of future impacts (NNC, IOM, PERG, TCPA, WIERC). Of particular concern was the build-up of long-lived highly toxic alpha-emitters in the environment, which over the long periods of time in question (hundreds of thousands of years in the case of plutonium) might find their way back to man in a manner not amenable to control. In addition, given the complexity of the environmental pathways and the uncertain basis of much of the biological data on transfer coefficients, there would always be the possibility of local concentrations, or of unusual habits, leading to high exposures. Thus a great many points were canvassed particularly where no data were given and gaps in monitoring were assumed by various parties. Whenever a suggestion was made in this direction, Parker frequently responded with the question, 'are you suggesting there is a hazard, or any cause for alarm?' In his report he treats certain cases at great length, commenting that parties sought to suggest there was cause for immediate concern. In all cases that we can trace, each party has been at pains to make this clear: that they were not suggesting an immediate health hazard, rather that trends were worrying, and that gaps in knowledge were sufficient in themselves for concern to be grounded (NNC 94.53D, 56F—H and IOM 38.11G—H, 93.60, 75C, 23.11—22).

4.5.78. Specific examples of gaps in the monitoring data (at least that presented at the Inquiry) were given by Taylor, where PERG had analysed sea-gull eggs and discovered an hitherto unsuspected pathway. PERG made it clear that they were not suggesting an immediate hazard, but it was noted that mussels which formed part of the pathway, and had at times been collected by locals, were now eschewed as the news of the eggs had been spread by the media. Parker stated that he was most concerned the public should not be alarmed. Bowen also indicated that IOM scallops, which constituted small industry on the island, had levels 40 times higher than Windscale fish and yet had not been monitored. He suggested that if basic standards for plutonium were to be altered by factors of 1000, as had been suggested by some authorities, the industry could be endangered (38.40, 60—63, 64B—C, 47.36, PERG 15, 15A, G63).

4.5.79. BNFL's response to these situations was to calculate the amount that a consumer would have to eat to reach the ICRP limits (assuming present standards were correct). PERG, NNC and IOM pointed out that this missed the point of the criticism. Parker devotes several paragraphs to exercises of this nature for IOM potatoes (a potential pathway noted by Bowen), scallops and local water supplies (a question by Urquhart which had been taken up by the press at the time). He concluded that the gaps in monitoring were not significant, and that current levels presented no immediate danger.

4.5.80. A particular object of criticism was the idea, central to control strategy, that there could be calculated a Limiting Environmental Capacity on the basis of the data available (19.40). For alpha-emitters, for example, MAFF had calculated that the LEC for plutonium was 310,000 Ci/a and for americium 250,000 Ci/a. Thus in their view three-quarters of a million curies of alpha could be released annually by BNFL without the ICRP limit being reached. The critical pathway for this LEC was held to be Porphyra and previous discharge control had been based on this assumption. The authorised limit of 4000 Ci/a had been raised to 6000 in 1970.

4.5.81. NNC argued that such a strategy was inadequate for nuclides of long half-lives that behaved in a conservative manner, and accumulated, as did plutonium in sediments whose behaviour was not predictable. This case was also made by Bowen for the IOM. NNC documented the use of Porphyra DWLs (the seaweed having been harvested and marketed in South Wales). It was argued that MAFF had altered the DWL as levels had risen and consumption patterns changed, but had also made questionable assumptions about those patterns. For example control had at one time been based on the failsafe principle that Cumbrian sea-weed could be eaten undiluted by weed from elsewhere. This principle had been dropped at the same time as levels had risen and that if it were to be reinstated then on a number of recent occasions Porphyra would have exceeded its DWL (41.79—80, 26.92—96, 70.44). Mitchell accepted this proposition (47.25).

4.5.82. The question of synergism was taken up by all parties concerned with monitoring but no evidence was available. It was agreed that a person regularly eating fish and shell-fish, walking on silt at Ravenglass, breathing resuspended plutonium and subject to gamma irradiation from contaminated sediments, would thus be subject to a synergism of pathways, but no figures or surveys were available relating to the habits of such individuals. The possibility of biological synergistic effects of radiation and other pollutants (and also of smoking) was canvassed by several parties. These questions are exemplified by the case of the IOM (37.20—21A—B, 93.67—68).

4.5.83. Taylor provided an example of a prediction by the controlling authorities that had not been born out in practice. Data submitted to UNSCEAR for 1977 (BNFL 115), by the NRPB gave an estimate for the maximum collective dose to be expected from the caesium discharges and consumption of Irish Sea fish in the light of measures taken to control the discharges. A figure of 6000 man rems had been quoted and was derived from estimates provided by MAFF. However, the latest MAFF report showed that in spite of a fall in caesium discharges, the levels in fish, or patterns of consumption, were such that the maximum predicted had been exceeded by

100%, at 12,000 man rems (96.8D, 51.9E—F). Taylor concluded that even for the best researched pathway of MAFF, prediction was not possible to within reasonable accuracy.

4.5.84. Several parties expressed concern at the very high projected emissions of krypton-85 (circa 15 million curies per annum). Taylor pointed out that the impact of Kr-85 was predicted from models only, and that these might have to be revised as more data became available. He gave an example of such a case of revised modelling from the studies of Bradwell nuclear power station (PERG 27). The CEGB had developed a model that predicted 200 mrad doses at the site boundary, whereas the NRPB model (data submitted to UNSCEAR, 1972), gave figures of 0.6 mrads (Bryant 51.14—16). In the light of the figures predicted by the German conceptual plan for a THORP (PERG 1), Taylor felt that dose figures should be recalculated for Windscale and validated where possible by measurement of the plume. Bowen made this same point, arguing that the plume might not disperse as readily as predicted (37.10A—B).

Protection of the Public: Ravenglass

4.5.85. We take Ravenglass as a special case as it embodied virtually all of the above issues in the dispute concerning the significance of higher levels of alpha-emitters than had been hitherto predicted by the authorities. As noted above (cf. 4.3.71, 72), the Ravenglass silt had been contaminated by plutonium since Windscale started discharges. Until the mid-1970s monitoring had been confined to the biota, in particular mussels, and to the silt itself. However, Hermiston gave figures for plutonium in air in the estuary which had not been published before and had resulted from measurements taken in 1976. His maximum value of 0.05 pCi/m^3 and a mean of 0.007 pCi/m^3 from a sample of 20 was indicative, he agreed, of a resuspension pathway, but did not constitute an 'external radiation hazard' (Hermiston, proof paras. 57—60).

4.5.86. This was also the view of MAFF who had instigated a series of air-monitors, using semi-quantitative 'Tacky Shades' (perhaps as a response to a television programme in which Bowen and Livingstone, who had been taking a professional interest in the accumulation of alpha-activity in sediments and had argued that remobilisation and resuspension were distinct possibilities). The monitors had found no evidence of elevated levels at the time (G 45, Laxen 69.30—41).

4.5.87. The views of BNFL and MAFF, that the elevated levels of plutonium did not constitute a hazard were strongly criticised on the following grounds: (1) A fundamental aspect of control was 'predictability'; the resuspension pathway had not been predicted by the authorities, and the LEC for plutonium had formerly been based on the levels of the nuclide in seaweed; no accurate predictions could now be made as to the future behaviour of the plutonium, and the possibility of Pu contaminated sediments building up in the estuary could not be ruled out (Bowen 37.37G–H, 36E, 63, 71–73). When asked about the potential problem by the MP, Robin Cook, the Secretary of State had replied that it would take 'a cataclysm' to remobilise the sediment and cause it to accumulate (37.16D–F). (2) If one accepted that Hermiston's figures and the ICRP standards were correct, then MAFF's LECs for plutonium at 310,000 Ci/a and americium at 250,000 Ci/a, would in this case lead to 3100% of ICRP, thus demonstrating the inadequacies of the LEC approach (Laxen 69.30–41). (3) Future behaviour would be affected by unpredictable environmental factors such as weather conditions and changes in wind and tide patterns (96.8–9). (4) The standards for calculating the dose to lung from inhaled insoluble plutonium particles were in doubt and if revised by a large factor could seriously alter the picture as far as health impact was concerned. Radford in a detailed proof on alpha standards and the situation at Ravenglass, held that present levels in air could lead to a 10–100% extra risk of cancer (from the measurements made by NRPB at the time of the Inquiry, see below), and up to 2500% on the maximum figures given by Hermiston (75.92). (5) In this latter respect NNC had argued that the assumptions made by the authorities with regard to the public exposure at Ravenglass were inadequate because the resuspended plutonium could travel the few hundred yards to the village on the estuary, it could there accumulate as dust and therefore it was appropriate to take the conservative assumption that people (including young children) were exposed continuously and not as Hermiston had assumed, for 300 hours per year (the time spent by the most exposed salmon fisherman on the silt). Laxen estimated that if the highest levels recorded by Hermiston were repeated for the whole year and were experienced in the village, then adults could be exposed to 12% ICRP (current standards) and children to 82% of the limit. (6) There were further questions concerning the relationship of NRPB and BNFL's DWL for soil, which was found to be close to the levels found in Ravenglass silt (Morley, 50.74–76, 64). There was doubt also about the standards for americium, which it was argued NRPB and BNFL underestimated (Thompson 68.2–4, 6–11). (7) Further doubts about the data base upon which the authorities calculated the behaviour of plutonium were expressed, and these were summarised by the IOM in a quote from a paper by Hetherington of MAFF, who had since resigned his post, where he stated that there was an 'appalling lack' of knowledge on the behaviour of these nuclides (IOM 86, and 93.75). Further documentation of the problem was provided by Urquhart (WIERC 16, 17, 18, 19, 21, 24 and 51, and 25.79D–E, 26.3–16).

  4.5.88. Parker's response to these criticisms was to order immediate sampling of Ravenglass air (48.1). It was pointed out at the time by CCC,

Table 30. Summary of the results of daily measurements of gross plutonium-239 and 240 and americium-241 activity concentrations in air at Ravenglass during the four-week period 26 August to 23 September 1977 and a comparison of the mean values with the maximum permissible concentrations in air recommended for members of the public by the ICRP for both soluble and insoluble forms of the nuclides.

Nuclide	ICRP, pCi m^{-3}		Northern sampling site				Southern sampling site			
	MPC$_a$ sol	MPC$_a$ insol	Measured pCi m^{-3}		Mean as % ICRP		Measured pCi m^{-3}		Mean as % ICRP	
			Mean	Range	MPC$_a$ sol	MPC$_a$ insol	Mean	Range	MPC$_a$ sol	MPC$_a$ insol
Pu-239 and 240	6×10^{-2}	1	2.2×10^{-4}	7.0×10^{-5} to 7.3×10^{-4}	0.37	0.022	6.1×10^{-5}	1.0×10^{-5} to 1.5×10^{-4}	0.10	0.006
Am-241	2×10^{-1}	4	1.8×10^{-4}	6.0×10^{-5} to 6.3×10^{-4}	0.09	0.005	3.7×10^{-5}	1.0×10^{-5} to 1.1×10^{-4}	0.019	0.001

Note that the measured activity concentrations in Oxfordshire from nuclear weapons fallout were approximately 2.3×10^{-5} pCi m^{-3} Pu-239 and 240 and 3.5×10^{-6} pCi m^{-3} Am-241 during the same period.

BNFL, NNC and PERG that no scientifically useful sampling could be done in the time available to the Inquiry (48.1–4); however, Parker insisted that some sampling be done, and high-volume air samplers were disposed at Ravenglass for a period of several weeks. The results are documented (G 50, 54, 55, 61, 67, 71), and we reproduce the summary from G 67 in table 30.

4.5.89. It was concluded by the NRPB that the figures confirmed 'elevated concentrations' at Ravenglass, that the temporal and spatial variations were not unexpected and underlined the need for prolonged observations if one wished to estimate annual dose to the public, and that no clear correlation with wind, weather, or tidal conditions was observed. From a comparison with ICRP recommendations, it was concluded that 'during the period of observation' there was no danger to the public in Ravenglass from airborne plutonium and americium (77.81–82, 88.37F–38D, 47).

4.5.90. In cross-examination of NRPB witnesses the doubts as to the data base for model of dose-risk were further expressed. In particular it was admitted that the chemical form of the plutonium was not known, neither was the mechanism of resuspension, nor the mode of sediment movement into the estuary (96.8–9, cf. Mitchell 47.11–17, 20–25, 42.9–24, 28–29). Radford commented, 'I find the situation ominous indeed' due to 'the continued discharge of alpha emitters from Windscale, the fact that high onshore winds which will convey even more plutonium and americium to homes, tracking of contaminated mud on fishermen's boots indoors, the likelihood that house dust at Ravenglass may already lead to higher exposures for women and children than would be expected from outdoor measurements' (75.94A–B).

4.5.91. Bowen had already commented, 'if one proceeds for a long period to neglect local inventory accumulations and then discovers that in fact there has been such a local accumulation, and that it has begun to be remobilised, one is in a situation where there is no immediate remedy or recourse whatever. The inventory is there, the remobilisation is taking place and there is no change which man can then make in his operations or in the system which will pull him back from the mess that he is in' (37.8B–C).

4.5.92. Parker had already stated at the outset that he saw no reason for anybody to be worried (48.1–4), and he reiterated this when the results of the first monitoring data came in—even though at that time an error showed the figures to be at 10% of ICRP. In his report he noted that Potts had not commented upon Ravenglass, and attached weight to this—stating that had Potts considered the situation alarming he would have said so. Laxen had already emphasised that Potts had not studied the problem at all (99.77). Parker also gave weight to Ellis who had declared himself not to be worried, but had made no submission on the matter (77.79–80). He maintained that NNC had submitted that there was cause for alarm (see note above, para.

4.5.77) and that NNC's statement that no assurance could be given until an extensive research programme had analysed the situation could have no weight attached to it. He also considered the determination shown by NNC, and Laxen in particular, as 'somewhat fanciful' (Parker Report, paras. 10.83−84). NRPB had stated that they did not believe the situation warranted an extensive sampling programme, and Parker was of the same opinion (para. 10.85).

Protection of the Public: Waste Management Strategies

(i) *Marine Discharges*

4.5.93. The principal concern of this section is with evidence on the availability of alternative methods of discharge control, rather than with the philosophy or decision-making procedures relating to their installation. In this context evidence was presented by PERG and NNC on alternative technologies (or aspects of conceptual flowsheets), and in addition the Company were asked to provide details of alternative THORP discharge control concepts available to them. BNFL had argued that through their partners in the European Consortium 'United Reprocessors' a technology-sharing agreement operated, and that they thus had access to the latest technology and experience. However, they presented no details of such alternatives.

4.5.94. Wynne, for NNC, was primarily concerned with the procedures of decision making regarding particular aspects of discharge control: for example, whether or not to install a salt-evaporator for Medium Active liquors (16.45G), or plutonium/americium recycle (20.43−45, 46E). It was stated (see later section), that such decisions were not made with the use of exactly quantified cost-benefit analysis, but that the conceptual flowsheet was designed to an overall environmental impact, i.e. within the authorisations and at the level of intent indicated above. The critique with regard to control technology centred upon this philosophy, but was illustrated with particular reference to the control of transuranics.

4.5.95. This dispute was illustrated by Clelland, who when questioned by Taylor on the control of plutonium discharges, and asked to accept that there was some controversy on the size of the permitted release, answered, 'not as far as I am concerned' (17.71G). PERG presented evidence of stricter discharge standards, tighter control intentions and use of alternative control technologies from German and French oxide reprocessing plants and/or conceptual flowsheets (17.73A−B, 20.5, 83−87, 15.36−38, 61−68, 30.17F, 50.86−88, 96.9, 10, 16, PERG 1, 78). This argument was also supported by IOM and Urquhart, who provided evidence of the French use of 'floc' treatment for plutonium discharges (93.86, WIERC 45).

4.5.96. Taylor also noted that the German plant did not intend to discharge tritium without some form of control technology, and that the Germans regarded the discharge of large amounts of tritium to the environment as undesirable on health grounds (PERG 1, 20.5, 83—87). It should be noted that BNFL intend discharging approx. 1 million curies of tritium to the marine environment, and that MAFF estimate the LEC at 500 million curies (BNFL 160).

4.5.97. Taylor also made the general point that the USSR did not allow transuranic discharges to the aquatic environment (PERG 27), and that the Germans' planned discharges were totalling less than 1 curie from liquid discharges—making the point that the technology for retention was available (96.16).

(ii) Aerial Discharges

4.5.98. The general point was made by a number of parties that BNFL's authorisation did not include quantitative limits on the discharge of gases to the atmosphere, rather they were required to operate to 'the best practicable means'. This procedure had been criticised by the RCEP. The IOM and CCC in their closing statements recommended that limits should be set, and Parker accepted this (para. 10.115).

4.5.99. Controversy centred upon the specific example of krypton-85 discharges, which BNFL did not intend to control. References to the potential effects of krypton were made by a number of parties, including the IOM. However, Spearing provided evidence on the availability of krypton control technology from the USA (JKS 9, 10, 40.50—52, 48—49). Urquhart also brought a witness from the US to testify to the potential environmental effects of krypton (Boeck 80.10). BNFL and NRPB maintained that Kr-85 was an international problem and that unilateral action would not have a great effect (BNFL 97), and that furthermore, the health impact would not justify action until the next century; however THORP had been designed to enable control technology to be installed should it be sufficiently developed and required (Warner, proof para. 55). Parker concluded that unilateral action by BNFL would be desirable once the technology was developed (para. 10.52).

Protection of the Public: Alternative Waste Management Programmes

4.5.100. The central argument with regard to THORP concerned the necessity of reprocessing oxide fuel at all when environmental impact of disposal routes was considered. FOE stated that US policy had changed due to the problem of fuel economics and proliferation implications, and that engineered storage of fuel was considered acceptable. FOE accepted that reprocessing was necessary for Magnox fuel. BNFL countered that oxide fuel could not be guaranteed free from corrosion problems as had affected Magnox storage, and that the disposal of unreprocessed fuel elements also presented an environmental hazard, possibly greater than that of the intended vitrified high active waste.

4.5.101.   The argument focussed on the following technical points: (1) *Dry storage of oxide fuel elements:* FOE argued that this was the preferred route, allowing cooling in vaults until such time as direct disposal of the elements could be engineered (Clelland 17.73D, Warner 29.3, proof suppl. 29.1–27, FOE 67, 84). Silsoe and Warner replied that dry-storage might take 20 years to engineer and that the Company had no plans for it (29.2F–4B, 98.55E, 56–60). (2) *Wet storage:* although this route was not specifically advocated, a number of parties called for delay (including FOE and TCPA) and argued that wet-storage, as at present, would not lead to serious problems. Tests undertaken at the Inspector's request (there having been no specific research programme so far), divulged problems with the storage in water of AGR fuel. Corrosion was evident and the integrity of the fuel could not be guaranteed beyond five years (Flowers 86.57–65, 86.82, 98.56–62). It was noted that THORP could not be expected to be commissioned before 1987, ten years hence, and that bottling of fuel elements to provide a barrier was now standard practice. Certain criticisms were made of these projections in the light of possible counter-measures such as anticorrosion treatment of the water, but FOE made the point that a major redesign of the AGR fuel elements ought to take place to prepare for the possibility that THORP would not be commissioned (92.72A–B).

4.5.102.   BNFL's main criticism of the engineered storage was that reprocessing was necessary for final disposal and thus could not be avoided. The principal arguments related to the plutonium content of spent fuel compared to vitrified waste, which represented a potential hazard due to criticality. In addition they maintained that spent fuel would lose its integrity more rapidly than glass and that leach rates of such elements as caesium would be greater, leading to higher potential environmental impact from underground disposal sites. This view was countered by FOE with alternative analyses of leach rates for glass and fuel pellets, and that plutonium was far more hazardous separated out and above ground (4.42, 17.28–30, 56E, 77, 80, 81–89, 27.71–81, 96, 92.65–72A–B, BNFL 9, para. 377, 186, UKAEA 1, CEGB 9). FOE's conclusions were that leach-rates for pellets were not significantly different from glass, although it was recognised that swelling and rupture due to chemical change might prove to expose more surface area to leaching than would be the case with glass undergoing fragmentation. FOE, however, accepted that some form of processing might be desirable, but stated that there should be no plutonium separation (for proliferation reasons), and if possible no environmental discharges (92.75).

4.5.103.   Parker concluded that extended wet storage was too risky because of the uncertain nature of AGR fuel, and that air-cooled vaults would take too long to engineer. He stated that it was not surprising that no research had been done on the longevity of AGR fuel under water because reprocessing had always been assumed (paras. 8.6−32). In addition he added that vitrification 'must' be developed for Magnox fuel, and he was confident that it would prove successful (see below), and hence provided a ready disposal route.

Protection of the Public: Site Selection

4.5.104. It was evident from the history of the site (BNFL proofs, see above) that Windscale had been chosen quickly and for primarily administrative reasons. As dealt with under the history of discharges the preliminary assessment of the sea-disposal option had taken place over a short period and with small quantities of radionuclides.

4.5.105. The IOM held that increased knowledge of the impact of discharges to the environment led to the conclusion that Windscale was not inherently suitable for reprocessing and that no further installations should be added at least until alternative sites had been systematically evaluated. Bowen listed the disadvantages of the Windscale site and enumerated the qualities of a more suitable one, which would include rapid dispersal of radionuclides to open sea (and not to an enclosed shallow basin as in the Irish Sea), and to a shoreline that was rocky or sandy and not characterised by silt and estuaries (37.3−4, 38.15−17, 48, 51B−F, 93.38−42, 71.69−103). The argument for alternative site evaluation was supported by the TCPA (see Stoel and Thirlwall proofs and cf. 6.2.7, 8−19), and CCC (98.34).

4.5.106. Parker concluded that the infrastructural advantages of the site, plus the accumulated knowledge on discharges, outweighed the case for consideration of alternatives, of which he noted, there was no legal requirement (paras. 14.5−10).

RESEARCH AND ORGANISATION

4.5.107. It was contended by IOM, NNC and PERG that in a number of areas there was insufficient research, or a lack of plurality of research, in the basic biological fields upon which many models of environmental impact depended. One problem particularly highlighted was the organisation of research within the UK with respect to the aquatic, terrestrial and atmospheric environment. In these fields responsibilities were divided: the Fisheries Research Laboratories (FRL) of MAFF were almost the only organisation active in marine research; the Agricultural Research Council had undertaken research into the effects of fallout, but no longer carried out extensive work, while the responsibility lay with the MAFF Food Science Division, which had no research capacity. The RCEP concluded that there was insufficient research in the atmospheric and terrestrial environments, and recommended that the Natural Environment Research Council (NERC), and the NRPB in consultation with ARC and MRC, should undertake such work (BNFL 9, paras. 231–238).

4.5.108. This problem was highlighted in a number of cases concerning the inter-relationship of basic standards, data on which those standards were based, and models of the dose to man. The most extensive example was that of the transuranics (cf. 4.5.85–92). Bowen, whose expertise was in the geochemistry of transuranics in the marine environment, indicated that the trend of recent research was to show that plutonium and americium were more mobile in the environment than had been predicted from laboratory studies. The reasons were that laboratory studies dealt primarily with simple chemical forms of the element, such as insoluble oxides or soluble citrates and nitrates. In the environment plutonium could form complex organic compounds whose behaviour was difficult to predict—in particular the transfer ratios from one organism to another, along food chains to man, might be quite different; in addition the gut-discrimination factors for man depended on the chemical form of the plutonium, and this was, in many cases, not known. It was therefore not safe to disregard some pathways as not worth measuring, and equally unsafe to assume gut-discrimination factors where humans were ingesting regular but minute quantities (37.21, 23G–24B, IOM 81, 86, 87). Thompson for NNC made the same critique (67.44–48). Under cross-examination, Mitchell of MAFF agreed that the data base upon which their models were based came from ICRP Report 2 (BNFL 82), and that these chemical forms were simple and not necessarily related to those in the environment (41.92–103, 42.2–9, NNC 41, 13). The RCEP made the same point in relation to sediment remobilisation—a response to evidence from Bowen at that time (BNFL 9, para. 353). Mitchell in answer to questions by Taylor admitted that they did not know the chemical forms of plutonium in organisms, nor in the atmosphere around Ravenglass, and he agreed that the discrimination factors used in their models for dose could be out by a factor of 100 (47.26–32).

4.5.109. The response of BNFL was to underline their obligation to use ICRP standards, but that in addition they had relied upon the NRPB report R44 on actinides in the environment (BNFL 113). This report was held to be totally inadequate by Bowen as most of the information did not relate to organisms in their natural environment.

4.5.110. It was noted by Taylor that both BNFL and MAFF regarded NRPB as an authoritative point of reference on matters of basic research, yet the NRPB research role on environmental matters relating to discharges was minimal. Indeed NRPB admitted that the authorising bodies did not come to them for advice (Morley 50.86−88). Harper for IOM and Wynne for NNC criticised the lack of integrated research between monitoring organisations and advisory bodies and the authorising departments (94.46E, 68−70, 93.76−85, Bowen 37−8B−C). Wynne also noted that the director of FRL, Mitchell, had not seen the new draft discharge limits until the Inquiry (G 28, 96.11−12).

4.5.111. The relationship of scientific research to the particular interests and outlook of the research worker and/or institute were stressed by Radford, Bowen and Wynne, in proof. Bowen, in particular, gave examples of 'straw houses' upon which theories were often based. He offered two speculations in relation to marine research, both of which he regarded as not disproved and therefore possible, but which went against current dogma: (i) was that americium-241 formed *in situ* might, because of its reactivity, behave differently than 'old' americium incorporated by sedimentation; (ii) that organisms might have evolved barriers to naturally occurring actinides (referring to the metabolic effort to exclude radium that he had often observed in aquatic organisms), but not to man-made nuclides such as plutonium, and that radiation from alpha-emitters might thus be biologically more significant than that from beta-emitters such as potassium-40 which also occurred naturally and within the body. He also regarded the possibility of a threshold for potassium-40 as a possibility because of the time available during the evolution of organisms for repair mechanisms to develop. Parker accords these two examples a whole paragraph, whilst not illustrating their context (para. 10.66).

4.5.112. The IOM, NNC and PERG concluded that research and organisation of control were inadequate, Wynne stating that it was 'seriously inadequate' and the IOM stating that they felt the development should be held up until the UK authorities had 'put their house in order' (94.68−70, 93.35, 48−49). The impression gained by these three critical organisations was that the institutional control of discharges was 'apparent' only, and due to inappropriate organisation, lack of co-ordination and interchange of knowledge. Both Wynne and Taylor also remarked on a general lack of awareness of modern environmental values (96.11).

4.5.113. Parker's conclusions on research are set out in paras. 10.127−129 of his report. He refers to the Government response to RCEP (BNFL 170, paras. 25, Annex A, para. 2, 11, 12). To which he had nothing to add. He held that there was sufficient independent research, and rejected any suggestions that research results had been kept secret by nuclear institutions or controlling bodies (10.129). However, the RCEP had stated that freedom to publish material was important, and the Government had stated such freedom existed, but in response to a question by Wynne, the MRC indicated that the Official Secrets Act did apply to the results of research work carried out in an official capacity (G 49).

4.5.114. The Government White Paper notes the RCEP views on gaps and lack of co-ordination, and states that future research will be co-ordinated by the Department of the Environment (not the NRPB as RCEP recommended), and that a review of control procedures would take place.

Case references
Ashworth 71.61−103; Boek 80.10; Bowen 37.3−4, 8, 10A−B, 16, 20−24, 36E, 37G−H and 71−73, 38.11−17 and 40−70; BNFL 98.55E and 56−60; Bryant 51.14−16; Clelland 17.28−30, 56, 73A−B and 77−89; CCC 98.34; Corbet 27.71−81 and 96; Ellis 77.79−80; FOE 92.65 and 75A−B; Flowers 86.57−65, 82 and 98; Hermiston 23.11−12, 25.79D−E, 26.3−16, 17−22, 37−42 and 92−96; Hookway 40.48−49, 50−52; IOM 93.35, 38−42, 48−49, 60, 67−68 and 75−86; Laxen 69.30−41, 70.44; Mitchell 41.79−80, 92−103, 42.2−29, 47.11−17, 20−32 and 36; Morley 50.64, 74−76, 86−88; Mummery 19.40 and 43, 20.43−46 and 83−87; NNC 94.46, 53D, 56F−H and 68−70, 90.77; NRPB 77.81−82, 88.37F−38D and 47; PERG 96.8−12, 16; Radford 75.92, 94A−B; Shortis 16.45G; Thompson 67.44−48, 68.2−4, 6−11; Warner 15.2−4, 36−38, 61−68, 29.3, 30.17F; Webb 51.9E−F

Proofs
Boek; Bowen; Clelland; Hermiston; Laxen; Stoel; Thompson (NNC); Thirlwall

Documents
BNFL 9, 82, 97, 113, 160, 170; CEGB 9; FOE 67, 84; G 28, 45, 50, 54, 55, 61, 67, 71; IOM 81, 86, 87; JKS 9, 10; NNC 13, 41; PERG 1, 15, 15A, 27, 78; UKAEA 1; WIERC 16, 17, 18, 19, 21, 24, 25, 31, 45, 51

ACCOUNTABILITY AND ACCEPTANCE OF CONTROL

Issue: Whether the system of control was sufficiently accountable and therefore acceptable to the public.

4.5.115. As noted in previous sections (cf. 4.5.9−47), doubts were cast on the openness of the authorities to a revision of standards. It was clear that some parties did not accept the controlling bodies' assessments of risks as derived from basic standards. The IOM Government and NNC in particular provided witnesses who testified that in their view the ICRP, the basis of UK standard setting, was not sufficiently accountable, being a self-perpetuating body (Radford, Bowen, Ichikawa). The dangers of this with regard to a bias toward research results resulting often from a perspective gained in the service of radiology or power generating industries were described in Radford's testimony. (It was noted that key UK members of ICRP, and NRPB were former employees of the UKAEA.)

4.5.116. Parker commented, 'I accept that a largely self-selected body may tend to perpetuate its own thinking, but it need not do so. . . .' He regarded the international scientific community as a sufficient safeguard (paras. 10.101–102).

4.5.117. In order that safeguards could be effective NNC had argued that the function of risk assessment and recommendation of limits should be separated. This was current US practice. Parker did not agree that this was desirable, feeling it would proliferate organisations. He felt ICRP were well placed to do both jobs and could 'bring together the appropriate experts' (paras. 10.106–108).

4.5.118. With regard to the critique of the NRPB and its links with the UKAEA (96.10–14), Parker agreed that the NRPB must be seen to be independent. He had no doubt that it was. He accepted that suspicion existed and recommended for this reason that an independent environmental interest be party to authorisations (paras. 10.110–111). He did not present any of the evidence on membership of NRPB nor of ICRP.

4.5.119. With regard to safety assessment, a similar critique concerning closed expert groups, links with industry and the exclusion of certain parameters from standards of safety assessment was made by PERG of the NII. Taylor quoted the RCEP commenting on the difficulty of acquiring expertise outside the industry (BNFL 9, paras. 292). Warner replied that many of the people who vetted his designs could be regarded as 'poachers turned gamekeepers' (15.74–76). Wynne also discovered in cross-examination of Niven that four of the six Radiochemical Inspectors, a director general of research and a superintendent chemical Inspector, were formerly with the UKAEA (46.30–32).

4.5.120. Thompson had stated in proof that as long as secrecy was necessary, then NII could not be seen to be doing an independent job, for there was no basis for open public assessment by critical experts. Parker drew much the same conclusions: that the NII were doing their job adequately, but that in this case security considerations were necessary. Widdicombe for WA called for an independent Inquiry into safety (91.10), and Thompson made the point that technologies which necessitated secrecy could not guarantee acceptability.

4.5.121. The role of the authorising bodies in setting discharges was questioned from the point of accountability. The IOM, PERG, NNC and LWSFJC all concluded that control was ineffectual in crusial respects and that a review should be instigated with thought to an independent outside body setting the limits (37.63–64, 29–32, 41–42, 65.65F–G, 93.52). Parker noted that reviews were under way in the light of the RCEP recommendations for a unified inspectorate and he agreed that this should be expedited. However, he did not feel that problems of competence or organisation necessitated this. He did agree that the control procedures were not sufficiently accountable in that the applicant alone had the right of appeal on discharge limits, and indeed that the local authority had no right to be consulted, let alone to appeal. He recommended that the local authority be given such statutory rights (paras. 10.109–117, 118–122). It was also noted that responsibility for control had moved entirely within the political aegis of the Secretary of State for the Environment in respect of discharges, and the HSE with regard to safety (reporting to two Secretaries of State, Environment and Employment).

4.5.122. In order to increase the accountability of the authorities both PERG and NNC had also argued for some degree of public participation in control. Parker regarded any participation other than by elected local authorities as 'unreal' (paras. 10.109–117). This matter is dealt with more fully under Democratic Accountability below.

Case references
Bowen 37.29–32, 41–42 and 63–64; IOM 93.52; Niven 46.30–32; PERG 96.10–14; Potts 65.64F–G; Warner 15.74–76

Proofs
Thompson (PERG); Wynne

INTENTIONS

Issue: Whether in the light of BNFL's past activities they were likely to meet their intentions with regard to environmental impact.

4.5.123. It was argued by IOM, NNC, PERG and WA that in the light of the past record of BNFL and the uncertainties of the future that the Company could not be relied upon to meet its intentions with regard to environmental impact. There were four strands to this argument: the likely efficacy of discharge control; general technical problems with the mechanics or chemistry of reprocessing; the predictability of the effects of discharges radioecology); and the technical and social problems surrounding an acceptable disposal of solid waste.

4.5.124. Thompson for NNC had argued that BNFL was intending to reprocess fuel that was five times more radioactive, contained 30–40 times more transuranics, with an alpha-activity four times that of previous fuels, and that when a similar jump had been made in the past from low to higher burn-up fuels, severe problems of control had been encountered (67.23–28,

34). In addition AGR fuel elements in water storage were already showing signs of cladding corrosion after four years (86.82), and this presaged similar problems to those experienced with Magnox in the ponds. Warner had countered this by describing the proposed bottling of fuel and the installation of a pond water treatment plant (proof para. 20). Radford argued that the new procedures should be tested on the Magnox plant for three years before THORP was given the go-ahead (75.68D–F).

4.5.125. In a detailed proof, Shorthouse had argued that BNFL were using inappropriate techniques for modelling the efficacy of the future THORP processes. He was concerned that the scale up was too great a leap and that the test rigs at 1/5000th normal scale were too small to give accurate or realistic predictions of performance. Warner had countered this by stating that earlier scale models had proven effective and that BNFL had the requisite experience (15.79–80). In addition, the vitrification programme had not yet been proven on an industrial scale and it was argued there might prove to be unforeseen problems (17.8–12).

4.5.126. Parker concluded that with regard to the intentions of control and technical feasibility, he was in no doubt BNFL could meet their intentions.

4.5.127. IOM, NNC and PERG argued that current knowledge of radio-ecology was imperfect, especially with regard to the build-up of alpha emitters in sediment and the estuarine environment. Such validation studies as had been carried out on other nuclides had already brought surprises, and there was sufficient evidence that plutonium may behave differently than predicted (94.40, 37.42–44). Hetherington, of MAFF, had stated that there was an appalling lack of knowledge in this area (IOM 98, para. 310). Parker declared that he was satisfied with the present state of knowledge and that in any case a large margin of safety existed; for example past errors with Magnox had led to a maximum of 44% of ICRP limits to the coastal fishing community, and that as the limit did not represent a danger limit (agreed by Bowen), the safety margin was adequate and even were ICRP limits to be exceeded this would not be great cause for worry (paras. 10.37–40).

4.5.128. With regard to waste disposal plans, several parties questioned the models and assumptions used for safe concepts, in particular disposal to granite (Tolstoy 64.66–71). Taylor argued that the most likely barriers to acceptable disposal sites was social not technical, and he gave evidence of the social structure and psychology of determined opposition to waste disposal in UK and Europe (Taylor, proof and 96.22–24, 25). Widdicombe regarded the disposal problem as 'fundamentally insoluble' for these reasons (91.39–46). On technical matters BNFL had stated that in their view it was 'widely accepted that disposal would become a reality' (Clelland 17.50G) and that it was 'inconceivable that it should not become successful'. Bowie, former chief geological consultant to the UKAEA at the Institute of Geological Sciences, was also of the opinion that disposal presented no problems (44.21–28). Clelland had admitted when questioned by Taylor that sociologists were not involved in the planning of waste management strategies.

4.5.129.   Parker concluded that on the matter of vitrification he had no doubts it would prove successful, and that indeed, disposal 'must' be successful on account of the present Magnox liquid wastes (para. 8.30). He did not refer to the social barriers.

Case references
Bowen 37.42–44; Clelland 17.8–12; Flowers 86.82; NNC 94.40; Radford 74.68D–F; Thompson 67.23–28 and 34; Warner 15.79–80

Proofs
Flowers; Shorthouse; Thompson (NNC); Warner

Documents
IOM 98

5. Conventional Planning Issues

5.1. SUITABILITY OF THE SITE IN RELATION TO THE NATIONAL PARK

Issue: Whether the proposed development was a suitable activity to have in an area of great natural beauty, and whether alternative sites had been considered with this in mind.

5.1.1. Glidewell, quoting from the agreed statement outlined the national, regional and local settings of the works, and the formal physical planning considerations of the site, including some detail of the actual planning permission applied for by BNFL.

5.1.2. Allday, under cross-examination by Dobry (6.35B–37D) on the planning criteria adopted in coming to the decision to apply for planning permission at Windscale, argued that the most important factor was that Windscale was already an established reprocessing and plutonium site. In response to questioning about what criteria would have been applied if Windscale did not already exist Allday itemised (a) distance from a built up area; (b) good building land preferably on the coast; (c) access to water; (d) access to a reliable electricity supply; (e) good communications. However he insisted that other sites had *not* been considered for the THORP development. Allday repeated this under cross-examination by Layfield. Scott under cross-examination by Dobry (11.48B–G) agreed that the matter of alternative sites to that at Windscale had not been discussed with local planning officials. He argued (11.45C–46D) that the activities at Windscale were less harmful to the National Park than most other industries, and considered the site itself to be striking rather than offensive when viewed from the Park, but did agree (11.43G) that THORP would ideally be sited on a rocky coast rather than a sandy one.

5.1.3. Hopper compared the area of the Windscale site with other areas in Cumbria, and on the Cumbrian coast, and argued that the area around Windscale was less attractive than most other areas.

5.1.4. Parker (para. 15.9(ii)) pointed out that there was no legal obligation for BNFL to consider and rule out alternatives, considered BNFL correct in thinking a consideration of alternative sites to be unnecessary, and (para. 14.10) concluded without hesitation that Windscale was the proper site for the development of THORP.

Case references
Allday 6.35B–37D, 9.37F–38A; CCC 79.6A–20H; Hopper 79.16A–F; Scott 11.43G, 45C–46D and 48B–G

Proof
Hopper, paras. 10.56–10.61

Document
CCC 34

5.2. AMENITY

Issue: Whether the proposed development could be constituted as an unreasonable visual intrusion upon the landscape.

5.2.1. Thirlwall criticised the agreed statement for underestimating the effect of the massing of the proposed development, arguing that it would have a major effect on views of the works.

5.2.2. Berry and with the aid of photographs (FLD 1–11) put forward the view that the proposed additional buildings would add materially to the bulk of the works, and would increase its appearance of massiveness in the rural landscape. He requested that due weight be given to the impact of the proposals on the National Park.

5.2.3. Baynes said that the Lake District Special Planning Board considered that the visual impact of the development could be accepted given the presence of the existing buildings.

5.2.4. Ashworth agreed with the inspector that the new buildings could not make the site more visually intrusive than it already was.

5.2.5. Glidewell said that the County Council felt that the proposed development would have no detrimental effect on views of the works, or the appearance of the countryside.

5.2.6. Parker (para. 14.16) remarked that he was satisfied that there was no case for refusing permission on grounds of visual impact, and that if permission was granted there was scope for improving the appearance of the site.

Case references
Ashworth 71.88C–89A; Baynes 81.69C–70B; Berry 91.73A–75B; CCC 32.16F–H; Thirlwall 80.44B–48E

Proofs
Baynes; Berry; Thirlwall paras. 16−22

Documents
CCC 34; FLD 1−11 inclusive (photographs of the site)

5.3. INFRASTRUCTURE

Issue: That the THORP development would put an unreasonable strain on the local authorities' abilities to provide essential infrastructure to cater for the likely immigrants, and reduced the opportunities for local inhabitants in obtaining, for example, council housing.

5.3.1. Liddle outlined the Department of Transport's responsibilities for highways and their current programme of local improvements. Glidewell (32.17A−G) asked the Inspector to draw to the attention of the relevant government departments the effects of expenditure on essential infrastructure relating to the development, on the county council's budget, and outlined (78.19C−20H) the projected housing needs for the new development on a village by village basis.

5.3.2. Rich outlined the current transport situation, including road and rail access, by provision and itemised the improvements planned in the event of the development proceeding.

5.3.3. Alexander outlined the North West Water Authorities' water supply and sewage disposal plans, and argued that demands for both services could quite adequately be catered for if BNFL's proposed development went ahead, subject to appropriate statutory authorisations.

5.3.4. Barratt in cross-examination of Alexander, questioned the adequacy of the sewerage provisions in the area.

5.3.5. Sefton described the Windscale transportation study detailing the proportions of Windscale workers using bus, train and car, and itemised improvements that could be made for each mode.

5.3.6. Dalgleish requested that the government make an early move on improvements to the coastal trunk road and called for the maintenance of both freight and passenger services on the west coast railway.

5.3.7. Donnison reiterated the provision of the agreed statement, indicating that accommodation had been based on a 50% local recruitment figure.

5.3.8. Hopper summarised the accommodation requirements. He concluded that the applicant's (BNFL's) could be satisfactorily met subject, in the case of local authority dwellings, to appropriate government finance. He also outlined the needs for public services and said that the detailed proposals were subject to an agreed statement under which BNFL would make financial contributions towards local highway improvements.

5.3.9. Denson, in his opening speech for Copeland Borough Council, said that the Council felt that the infrastructural provision at present was totally inadequate for its likely increased needs in the event of THORP proceeding, particularly in relation to highways, sewerage, and sewage disposal.

5.3.10. Bailey said that the Borough Council policy was to ensure that developments associated with Windscale take place in the most suitable locations and at the appropriate rate, and that these developments were consequent upon necessary improvements in infrastructural facilities relating to sewage disposal and communications (see above).

5.3.11. Swift outlined the historical problems of infrastructure provision in Copeland BC and itemised the areas of particular concern relating to sewage disposal and highway upgrading.

5.3.12. Baynes said that the Lake District Special Planning Board felt that increases in traffic as a consequence of the development proceeding would not have any significant bearing on the National Park.

5.3.13. Parker (para. 14.28) made no recommendation with respect to the maintenance or improvements of the rail passenger service, nor (para. 14.29) on the assurance of the Department of Transport that 'they were examining the possibility of small local improvements throughout the length of the A.595' as regards road improvements, and concluded (para. 14.30) that road and rail traffic problems could be dealt with and constituted no grounds for refusing permission. As regards housing needs he remarked (para. 14.33) that he had heard no evidence which led him to reject the views of the local authorities that they could meet any housing needs, and accordingly accepted them. He also concluded (paras. 14.34—35) that the position as regards water supplies and sewerage and sewage treatment gave no ground for refusing planning permission.

Case references
Alexander 79.57A—61C; Bailey 81.54C—57A; Baynes 91.70C; CBC 81.44A—45E; CCC 32.17A—G, 78.19C—20H; Dalgleish 79.97F—G. Donnison 81.11B—12F; Liddle 79.50G—51C; Hopper 79.21F—27A; Rich 78.21A—25D; Sefton 79.72E—78F; Swift 81.64E—67D; TCPA 79.63H—67F

Proofs
Alexander paras. 3—16; Bailey paras. 5.1—6.9; Baynes; Dalgleish para. 8.5; Donnison paras. 7.6.1—7.6.6.; Hopper paras. 12.1—12.61; Sefton paras. 2.1—4.6; Swift paras. 1.11—2.11

Statement
Liddle

Documents
BNFL 293; CCC 34, 35

5.4. LOCAL EMPLOYMENT AND TRAINING SCHEMES

Issue: Whether or not the THORP development would create significant numbers of jobs for local people; what proportion of the jobs created would actually be filled by local people; and whether a large industrial complex of this kind had other adverse effects on the local employment pattern.

5.4.1. Glidewell in his opening statement for Cumbria County Council (32.15D—16E) said that the County Council felt that a major expansion at Windscale gave a chance to improve the employment prospects for the existing inhabitants of West Cumbria, and would be socially and economically beneficial to West Cumbria. He outlined (78.17A—19B) the current employment levels at the Windscale works and the estimated amount of future local recruitment for the works, as well as producing unemployment and migration statistics.

5.4.2. Silsoe, in his opening statement, said that the development would bring a sizeable number of stable jobs to a special development area, and expressed the hope that as many as possible of these jobs would be taken up by local people.

5.4.3. Donnison outlined the employment background to the application, the proportion of local people with various skills that were expected to be employed, and the general economic problems of this area: population loss through migration, and the high level of unemployment.

5.4.4. Hopper did likewise, and estimated possible future employment trends with and without THORP, including the likely proportion of future local employment.

5.4.5. Dalgleish requested on behalf of the County Council that the government provide special financial assistance to enable school leavers and unemployed youths in the area to be trained, or retrained, prior to the opening of the new facility.

5.4.6. Denson, in his opening statement, agreed that there were employment disadvantages as well as advantages in having the BNFL expansion, but felt that the potential long-term benefits in stemming outward migration outweighed the disadvantages.

5.4.7. Bailey outlined the likely job generation of the proposed development, and admitted that the influx of working wives of scientific staff could worsen employment problems in the service sector.

5.4.8. This was a point taken up by Kidwell (11.77H–78C) who argued that the influx of wives of the new workers could increase competition in the female sector of the labour market. He further drew Scott's attention (11.61G–73D) citing FOE 80 and CBC 1, Appendix I, as evidence, that while the number of jobs available in the Whitehaven area increased between 1955 and 1975, so did the unemployment levels.

5.4.9. Bartlett cited BNFL 294 in response to questions by Kidwell on the proportion of vacancies that would be filled by local unemployed people.

5.4.10. Scott argued that the THORP development would be of some assistance in reducing unemployment locally, and estimated that 50% of those employed in THORP would be recruited locally.

5.4.11. Elliot argued that nuclear power technology in general, and the proposed expansion in particular, represented a very poor investment in terms of creating jobs compared with viable alternative energy generating and conserving technologies, and outlined some estimates of the job creating potential of alternative energy systems as worked out in the United States and called for more research on the potentialities in the British context.

5.4.12. Hall pointed to the relatively short-term nature of the development in terms of employment security, and to an example of the use of regional incentives to create jobs in the area for less than £1000 each.

5.4.13. Higham argued that the high proportion of the local male population employed by BNFL had a detrimental effect on small local firms attempting to attract skilled labour, and foresaw that these difficulties would increase if the expansion were allowed to proceed. Bartlett outlined some alternative projects for the West Cumbria area, using local materials and manpower, and she argued that investment in these alternative industrial projects would create more jobs for less capital investment than would THORP.

5.4.14. Parker (para. 14.23) concluded that the effect of the development on employment would be beneficial, albeit not great and therefore rejected the argument that THORP would have an adverse effect upon employment. He also (paras. 14.24(i)–(iii)) rejected the contention that BNFL's expansion would cause shortages of skilled labour for other firms in the area, that it was economically unhealthy for West Cumbria to have such a dominant employer, and that the capital cost of each job created was too high, commenting on this last point that the argument appeared to be divorced from reality as there was no evidence that were THORP to be refused, money would be forthcoming to create other jobs.

Case references
Bailey 91.51E–53A; BNFL 1.10D, 3.31B; Bartlett 82.31E–35A; CBC 81.40A–42C; CCC 32.15D–16E, 78.17A–19B; Dalgleish 79.97E–F; Elliot 92.7A–11E; FOE 11.61G–73D and 77H–78C; Hall 80.92F–93H; Higham 71.10E–11B; Hopper 79.16G–21E; Scott 11.61G–73D

Proofs
Bailey paras. 4.1–4.7; Bartlett paras. 4.12–4.20; Dalgleish para. 8.5; Donnison paras. 7.5.2–7.5.24; Elliot paras. 2.1–2.25; Hall paras. 5.13–5.15; Higham paras. 25–26; Hopper paras. 11.1–11.58

Documents
BNFL 28, 294; CBC 1; CCC 2, 28, 34, 35; FOE 80; TCPA 97, 100

5.5. ECONOMIC EFFECTS

Issue: Whether, in the event of the development proceeding, special allowance should be made by central government for the financing of essential infrastructural improvements that would have to be made by the local authorities concerned.

5.5.1. Higham argued that the THORP development would cause further demands on Copeland Borough Council's finances, particularly in regard to housing, with a resulting worsening in services.

5.5.2. Dalgleish requested on behalf of the County Council, that the government in administering grants to, and loan sanctions within the area, both under general legislation and under Special Development Area powers, treat monies expended in meeting the applicant's infrastructure needs as additional priorities to, rather than part of, monies required for normal development in the area.

5.5.3. Parker (para. 14.37) noted the argument, but concluded that none of the evidence suggested that THORP would be in any way exceptional in respect of the numbers of jobs not filled by local people, and therefore saw no case for exceptional treatment as regards expenditure on infrastructure etc.

Case references
Higham 71.8H–10D; Dalgleish 79.97A–B

Proofs
Higham paras. 21–24; Dalgleish para. 8.5

6. Democratic Accountability

6.1. PARTICIPATION AND ACCOUNTABILITY

6.1.1. There were many references to participation of the public in the decision-making procedures concerning the control of technology in general, nuclear matters in particular, and also on matters of energy policy. On matters concerning the acceptability of discharge standards and safety evaluation, NNC and PERG put forward detailed proposals for increasing participation, accountability and hence acceptability. In other areas criticism was more general.

Policy

6.1.2. Government witnesses were by tradition not asked to comment on policy matters. The latest energy policy consultative document was presented to the Inquiry (G 71). Dudman, however, did press the matter on the question of reprocessing policy, to receive the answer that it had always been policy to reprocess but that such policy was in abeyance pending the results of the Inquiry (44.2—15). Jones (DEn) felt that the only legitimate form of participation was via elected representatives in Parliament and the local authority, CCC, also submitted that this was the case, CCC having no right or duty to consider 'need', nor health and safety matters (32.19, 75—76, see also Herzig 44.19—20).

6.1.3. Taylor outlined the divisions that existed in society on the questions of energy policy. He gave sociological evidence of the relationship of political philosophies of decentralisation and low-energy ecological ideals to perceptions of risk. He outlined the implications for nuclear power and acceptable energy strategies. He concluded that society was divided and that nuclear expansion, symbolised in the public mind by the THORP decision, would emphasise these divisions and create conflict. He indicated that for centralised political institutions the decision to keep options open and appease powerful industrial interests geared to large centralised power generation systems was too readily seen as a reasonable response, whereas in fact it was likely to create conflict. He questioned whether 'majority' rule was the correct yardstick when minority groups (barely a minority in most cases) felt strongly enough to consider civil disobedience (89.90—89, 90.30—69 and PERG 35, 75). He provided detailed bibliographies and extracts of the work of the IAEA and continent governments on the roots of the nuclear controversy (Taylor proof).

6.1.4.   Parker could not see how this material was relevant to his decision (90.69—72).

Standards and Control

6.1.5.   It was submitted by PERG that on the question of discharge standards and decisions relating to control technology there was a legal requirement on the part of the authorities to cater for some form of participation. Taylor argued that ICRP 22 (BNFL 149) had stated that 'human, aesthetic and other environmental factors' (known as 'intangibles' for the purposes of cost-benefit analysis) should be taken into consideration in setting standards (96.9C—E). He produced evidence that MAFF had considered this recommendation in a technical report (PERG 7) where monetary costs had been sought for cost-benefit analysis (47.38—40). Wynne made the same points on the question of optimisation studies for discharge control technology (42.2—4, 46.88—89, 20.51—57, 94.38). It was argued that such decisions in order to be acceptable must involve the widest participation, in particular of those people suffering the detrimental costs, and that the whole procedure should be open to critical scrutiny (46.27—39, 69.70—87).

6.1.6.   Wynne gave examples of the range of values that had been considered by the authorities. In the UK, for example, the detrimental costs of caesium control had been estimated at £13 million for the UK population and £23 million for Europe by MAFF, whereas an equivalent exercise on US standards would produce £150 and £264 million respectively. The use of economic data for lives lost, ill-health and degradation of the environment was in itself questionable, but at least the matter should be open. In the case of MAFF the technical memo on cost-benefit application had come to light late in the Inquiry after detailed questioning (G 64A).

6.1.7.   The matter was all the more important, argued Wynne, because the authorities were now advocating 'cut-off' doses which would severely alter the calculation of detriment and thus favour the industry's approach to pollution control. He felt that the NRPB/ICRP present approach was directly opposite to that of the USA, and that their statements that the linear dose response hypothesis might overestimate low dose effects were 'unbecoming' (45.80D, 76—83, 94.41—42).

6.1.8.   Wynne contrasted the assumptions of NRPB/ICRP in this way: 'it seems to me to be perfectly rational for human beings to insist that risks which are imposed upon them in a relatively involuntary way by authorities of a centralised kind, . . . should be less than those risks which they are prepared to accept for themselves in an individual situation of choice' (70.12D).

6.1.9. Glidewell for CCC took up the problem of participation in discharge control, arguing that this could only take place via elected bodies, but that it was not the responsibility of the CCC. However, he stated, 'the body which does at the moment make the decision has no democratic accountability at all. I am wholly with Dr Wynne on that'.

6.1.10. Hall called for a greater level of public participation in decision making in general and with regard to energy policy in particular. He also criticised the decision making procedure in planning applications such as the one under consideration (the decision could have been made by the planning committee of the County Council), and suggested that Parliament should be involved in a formal way in such major decisions through the medium of a select committee.

6.1.11. The problem of acceptability and groups in society suffering detriment was further exemplified by Potts for LWSFJC. He stated that the fishing interests were caught on the horns of a dilemma: if they say there is a problem, then they suffer damage because people will not buy fish (see PERG 35), if they say nothing there would be little chance of rectifying the problem (65.54). He agreed that the individual risk was insignificant (65.55F), but went on, 'we are condemning 30–50 or so people to genetic damage and a few to death due to cancer', this was 'suffering and human life wasted' and 'if these casualties were concentrated at Windscale in the present generation there would be an outcry'. He maintained that 'we have come very close to the situation in which it would be necessary to close parts of the Irish Sea to fishing'. He noted that the Cumbrian Sea Fisheries Committee had not put in an objection and supposed that this was due to a dependence on CCC (65.43–54). With regard to the future setting of limits, he would prefer to see an 'outside' body carry out the work (64.64F–G).

6.1.12. On the question of the openness of the system, Bowen also warned that unless the bases of decisions were made clear, the public might well refuse nuclear expansion (37.1–3, 26G–H).

6.1.13. Further points in the same vein with regard to acceptability were made by FOE-WC (94.16). WA pointed out that one German State, Nordrein Westfalen had decided against nuclear power until acceptable waste disposal had been demonstrated (17.65). The British Council of Churches concluded that the risks of the reprocessing option as displayed at the Inquiry, were too great (94.18).

6.1.14. On matters of safety from accidents, a similar series of points were made by Thompson in proof for PERG. He argued that no fully open safety study of the kind prepared for draft circulation in the USA had been done for any nuclear installation in the UK. This level of secrecy meant that acceptability was meaningless, because the public could not know exactly what risks they were being asked to accept, or indeed had been accepted on their behalf. He further pointed out that the Farmer methodology, by which the magnitude of risk and target probability was judged, had never been put to Parliament (45.41–42D, 56, 96.20).

6.1.15. The response of the various officials to these questions was in most cases characterised by the view that the present system of expert committees and bodies was adequately accountable to the various ministers concerned. Webb, for NRPB, was of the opinion that to involve outside interests in standards or control was 'impractical', but he admitted it had not been tried (51.12). Fremlin for CCC regarded the setting of 'official limits' as necessary, in view of likely public reactions to detailed knowledge (36.46–70).

6.1.16. Parker held that 'much talk of public participation' was 'unreal' (paras. 10.109–117). The only sense in which it could be real was through the elected bodies. He took up Glidewell's point about the lack of representation of the Local Authority in the process of discharge control authorisations, and held that they should have the right to representation and to appeal (paras. 10.118–122). He also recommended that in order to increase the acceptability of decisions on standards and control 'some wholly independent and responsible person or body with environmental interests' should be included in the advisory process (para. 10.111).

Case references
Bowen 37.1–3 and 26G–H; BCC 94.18; Charlesworth 45.41–42D and 56; Clelland 17.65; CCC 32.19 and 75–76; Dunster 45.76–83; Fremlin 36.46–70; FOE-WC 94.16; Hall 80.91F–G; Herzig 44.19–20; Mitchell 46.88–89, 47.2–4 and 38–40; Mummery 20.51–57; Niven 46.27–39; NNC 94.38 and 41–42; PERG 96.20; Potts 65.43–54, 55F and 64F–G; Taylor 89.80–89, 90.30–69; Webb 51.12; Wynne 69.70–89

Proofs
Hall; Taylor; Wynne

Documents
BNFL 149; G 64A, 71; PERG 7, 35, 75

6.2. THE INQUIRY PROCEDURE

Issue: Whether the Inquiry procedure was unsatisfactory and that therefore no decision should be made.

6.2.1. A number of parties complained of difficulties during the course of the Inquiry. These difficulties we divide here into those in the nature of complaints arising from the pressures of timing or finance rather than with the structure of the Inquiry itself, and those more fundamental to the form the Inquiry took. WA submitted that the interval between the announcement of the Inquiry and the start of proceedings had been too short for adequate preparation. TCPA drew attention to the pace of the proceedings themselves, Barratt remarking that there was little time to sit down and study figures (90.76E). Urquhart also remarked on the difficulties of travelling and stress (27.53, 29.41) and had problems following up cross-examination. Requests for a break in the proceedings were turned down (16.68, 73−81, 22.53).

6.2.2. Parker remarked that submissions that the Inquiry should have sat at a more central place showed less regard for the participation of those most affected. However he did note the heavy financial burden on those national organisations who tried to maintain a full-time presence. He suggested that difficulties over timing and presentation of cases were occasioned by participants taking holidays.

6.2.3. More fundamental points were made concerning the form of the proceedings and the nature of evidence. For example both PERG and NNC, who dealt with health and safety matters in detail, found themselves submerged in documentation with little time to digest contents. PERG in particular was receiving crucial answers to questions and the results of analyses, up until the 80th and 90th days (e.g. TIRION requests, 20.84, 21.105, 21.9C−E, 22.1−2, 26.51 and BNFL 280, 282, 284, 299, 303, 308). Thompson, for PERG, outlined the form of safety study that would be necessary for full public appreciation of the risks.

6.2.4. A member of the public interrupted the proceedings and proffered the view that the Inquiry was a 'farce', and that the decision had already been taken (45.73).

6.2.5. These, and other requests relating to evaluation of the design from the safety or health viewpoint, as well as security, were not reason enough to postpone a decision in principle, argued Parker, for THORP was an outline application only, and doubtless the appropriate bodies would ensure that the necessary approved standards were met before the complete design was given the go-ahead.

6.2.6. Although it was argued by many parties that THORP should be viewed in the context of the potential development of the FBR (a point disputed by Parker), the Inquiry was restricted to the field of reprocessing because a further Inquiry had been promised on the FBR itself (42.77, 45.93−96, 46.9−14).

6.2.7. Hall, for the TCPA, considered that the issues raised at the Inquiry would have been better covered through the setting up of a Planning

Inquiry Commission, as constituted under Section 48 of the 1971 Town and Country Planning Act (80.87F). Parker considered that there might have been considerable force in the suggestion had specific alternative sites been seriously suggested, and had there been no existing facilities at Windscale and no store of information covering discharges from it. In the circumstances he saw little merit in the submission (para. 15.13).

Environmental Impact Statements:

Issue: Whether the proposal could be adequately assessed without a formal environmental impact statement.

6.2.8. Barratt argued that the TCPA's case for an environmental impact analysis was that the preparation of one before the Inquiry would have helped to identify particular areas of dispute and difficulty.

6.2.9. Stoel outlined the environmental impact statement procedure as used in the United States, and instanced the types of projects they were applied to, pointed out that EISs are prepared prior to the construction of all new nuclear power stations, outlined the generic analysis used in the case of nuclear power stations, and stated that the environmental impacts of specific facilities were also required. He also outlined the areas that an EIS for Windscale would discuss, and concluded that an EIA would result in a net saving of resources, and a more effective way of eliciting and testing the facts relevant to the decision.

6.2.10. Thirlwall itemised two types of EIA (GESMO type and Barnwell type) and outlined the essential features of such analyses (systematic identification of the impacts of the proposal; selection for detailed examination of those potential impacts which in their magnitude or importance could influence the planning decision; a detailed study of each selected impact, which would include a survey of the relevant characteristics of the existing environment; the preparation of a comprehensive report dealing with the whole of the EIA which had been carried out). He concluded that a sufficiently penetrating and comprehensive study had not been undertaken having regard to the scale of the development and the demands which it would make on the area.

6.2.11. Dobry, during cross-examination of Allday (6.51F–54E) drew attention to the Flowers Report para. 523 which suggested the use of EIAs, and recommended their use in this case. He also itemised the IOM's reasons for objecting to the application on planning grounds (6.54F–62E).

6.2.12. The RCEP (paras. 523–524) concluded that an EIA, similar to the American EISs should be prepared by the proponents of a nuclear scheme, the analysis not being confined to the effects of the first stage of the development but following through to the furthest point which current knowledge could attain. Its aims would be to enhance public understanding of the development in question, and to be part of a decision-making machinery that was explicitly political in its process.

6.2.13. Ashworth, while conceding that as the Inquiry was not a Planning Inquiry Commission and was therefore not obliged to look at alternative sites, felt that it was unsatisfactory that such a procedure had not been adopted, at least by BNFL, and that as a consequence the Windscale site had not been demonstrated to be either wholly appropriate *or* the *least appropriate* for the THORP development (see also 4.5.104–106).

6.2.14. Glidewell in cross-examination of Ashworth (71.87A–E) disagreed with his interpretation, and suggested that the right test to be applied was whether the site proposed was an unsuitable site or not.

6.2.15. Parker (para. 14.9) pointed out that there was no legal requirement for an EIA, and said that he was satisfied that all matters which might have been included in an EIA were properly investigated at the Inquiry, and that absence of one was no grounds for refusing permission. He did however accept that had there been such an analysis it might have saved time at the Inquiry, and thought it worth considering whether in particular cases in the future an applicant or planning authority should be required to prepare such an analysis.

Financial Disparity

Issue: Whether the marked disparity in financial resources between applicants and objectors materially affected the conduct or outcomes of this or possible future inquiries.

6.2.16. Kidwell drew attention to the adverse effects on the finances of objector groups at the anticipated length of the Inquiry.

6.2.17. Hall drew attention to the discrepancy in financial resources between the applicants and objectors and called for financial assistance to objectors at public inquiries in future. He also argued that such funding would improve the level of decision-making at future inquiries.

6.2.18. Parker (para. 15.8) noted the disparity but concluded that in this case the financial resources available to objectors had been adequate for the purposes of making a decision on the application. He noted (para. 15.9) that a number of objectors doubt that they would be able to repeat the exercise at a future inquiry and while drawing attention to this made no recommendation.

Case references
Allday 6.52F–54E and 54F–62E; Ashworth 71.70A–74B; CCC 71.87A–E; FOE 16.68F–69D; Hall 86.87 and 94D–H; IOM 6.52F–54E and 54F–62E; Stoel 77.2A–6H; TCPA 77.2F–3B, 90.76E; Thirlwall 80.36C–61A

Proofs
Ashworth paras. 2.1–6.3; Stoel paras. 2–21; Thirlwall paras. 5–41

Documents
BNFL 9; IOM 105, 106; TCPA 68, 69, 70, 80, 98, 99, 106

Appendix 1. Inquiry Transcript Index

British Nuclear Fuels Ltd.
(Counsels: Silsoe, Bartlett)

Opening Statement	1, 2, 3
Closing Statement	98, 99, 100

Statements

Silsoe	10, 30, 31, 45, 48, 59, 90, 93
Bartlett	67, 71, 73, 85, 86, 88, 91

Witnesses

Allday	4–10
Avery	30, 31, 56
Clelland	16, 17
Corbet	17, 27, 28
Donaghue	23, 24, 25
Doran	13–16
Hermiston	22, 23, 25, 26, 27
Hopper	79
Milne	27
Mummery	17–21
Schofield	21, 22
Scott	10, 11, 78, 79, 97
Shortis	12, 13, 15, 16
Smith	16
Warner	12, 14, 15, 29, 30
Williamson	27
Wilson	27

British Council Of Churches
(Gosling)

Witnesses

Ecclestone	70
Gosling	70, 94
Postlethwaite	70

Central Electricity Generating Board
(Fitzgerald)

Opening Statement	40
Closing Statement	96

Witness

Wright	40, 41

Copeland (Denson)

Opening Statement	81
Closing Statement	97

Witnesses

Bailey	81
Battersby	85
Swift	81

Cumbria (Glidewell, Rich)

Opening Statement	32
Closing Statement	97, 98
Applications	46
Statements	21, 24, 32, 41, 49, 64, 78, 79, 85, 92, 93, 94

Witnesses
Alexander	79
Dalglish	79, 80
Donnison	81
Fremlin	32, 33, 34, 36, 38, 39
Murray	81
Naylor	32

Cumbria Naturalists Trust

Witness
Halliday	70

Electrical, Electronic, Telecommunications and Plumbing Union

Witness
Adams	66

Friends of the Earth
(Kidwell, Thorold)

Opening Statement	51
Closing Statement	93
Applications	16, 17, 56
Submissions	44
Statements	10

Witnesses
Chapman	56, 57
Leach	54, 55
Patterson	51, 52, 53
Wohlstetter	58, 59

Friends of the Earth—West Cumbria
(Haworth)

Opening Statement	60
Closing Statement	94

Witnesses
Bainbridge	60
Corkhill	60
Dixon	60, 62
Haworth	60
Jones	60
Mcleod	60
Norman	60

Friends of the Lake District

Witness
Berry	81

Isle of man (Dobry, Harper)

Opening Statement	71
Closing Statement	93
Application	16, 41, 48

Witnesses
Ashworth	71
Bowen	37, 38
Bowers	71
Quayle	71

Justice (Widdicombe)

Opening Statement	83
Closing Statement	92

Witness
Sieghart	83

Lake District Special Planning Board
(Robinson)

Witness
Baynes	81

Lancashire and Western Sea Fisheries Joint Committee

Witness
Potts	65

National Council for Civil Liberties
(Blom-Cooper, Robertson)

Opening Statement 83
Closing Statement 95

Witness
Grove-White 83

National Peace Council

Applications 59, 73

Witness
Oakes 61, 96

Natural Resources Defence Council
(Thorold)

Witness
Cochran 63

Network for Nuclear Concern
(Wynne, Laxen)

Opening Statement 67
Closing Statement 94
Statements 10, 75, 81, 90, 99
Applications 87

Witnesses
Ichikawa 75
Laxen 69
Radford 75, 76
Thompson 67, 68
Wynne 69, 70

Political Ecology Research Group
(Taylor, Thompson)

Opening Statement 89
Closing Statement 96
Application 45
Statements 48, 73

Witnesses
Taylor 90
Thompson 89, 90

Ridgeway Consultants (Little)

Opening Statement 47
Closing Statement 97
Applications 44, 92
Statements 47, 83

Witnesses
Bowie 44
Fletcher 71
Greenhalgh 47
Little 71, 73
Macdonald 47, 48
Rippon 47, 48

Scottish Campaign to Resist the Atomic Menace

Witness
Paulin 65

Socialist Environment and Resources Association
(Sedley, George)

Opening Statement 82
Closing Statement 82

Witnesses
Bartlett 82
Elliott 82
Lewis 82

Society for Environmental Improvement
(Tyme)

Opening Statement 35
Closing Statement 95
Applications 16, 61, 85, 86

Witnesses

Armstead	86
Armstrong-Evans	86
Evans	86
Hall	35, 36
Leach	35
Musgrove	35, 66
Page	35, 66
Randall	86
Salter	35
Scargill	35, 66
Spencer	86
de Turville	86
Wilson	35
Wilson (Lord)	35, 36

Society of Friends (Keswick)
(Spearing)

Statement	39

Witnesses

Spearing (Dr.)	86, 88, 97
Spearing	88

(Dr. Spearing also appeared as an individual objector)

South of Scotland Electricity Board
(Edward, Macfadyen)

Witness

Tweedy	41

Town and Country Planning Association
(Layfield, Barratt)

Opening Statement	73, 74
Closing Statement	95
Submission	32
Statement	79

Witnesses

Ellis	77
Hall	88
Kneale	74, 90
Odell	76
Rotblat	74, 75
Stewart	74, 90
Stoel	77
Thirlwall	80

Uranium Institute (Glidewell)

Witness

Price	52

Windscale Appeal
(Widdicombe, (Alesbury)

Opening Statement	61
Closing Statement	91, 92
Applications	98
Statements	10, 32, 51, 54, 55, 85, 99

Witnesses

Atherley	84
Blackith	61, 62, 63
Coates	64
Davoll	61
Goldsmith	85
Jenkins	64
Pedlar	84
Shorthouse	84
Sweet	65, 77
Tolstoy	64
Wakstein	84, 85

Windscale Inquiry Equal Rights Committee
(Urquhart)

Opening Statement	80
Closing Statement	94

Witness

Boeck	80
Urquhart	89

Individuals

Boles	100
Chivall	70
Dalton	96, 97
Dudman	61, 93
Fish	54, 95

Witnesses

Beckett	54
Scott	54
Hatton	84
Henderson	60
Hillier-Fry	72
Higham	71, 95
Holden	71
Jones	78
Miller	98
Richardson	70
Robertson	48
Rosenthal	45
Sly	72
Spearing	88, 91
Stredder	85
Tosswill	60
Tremlett	69
Wadsworth	88

Government Departments and Agencies

Ministry of Agriculture, Fisheries and Food

Carr	47
Mitchell	41, 46, 47
Small	41

Department of Energy

Herzig	42, 43, 44
Jones	42, 43

Department of the Environment

Hookway	46
Niven	45, 46

National Radiological Protection Board

Bryant	51
Dolphin	49, 50
McLean	49
Morley	50
O'Riordan	77
Shaw	88
Webb	51

Nuclear Installations Inspectorate

Charlesworth	44, 45
Dunster	44, 45

Department of Transport

Liddle	79
Wilson	48

United Kingdom Atomic Energy Authority

Farmer	52
Flowers	86

Appendix 2. A Note on British Planning Inquiry Procedure, and Its Application at the Windscale Inquiry

The Windscale Inquiry was held as an ordinary public local inquiry under Section 35 of the Town and Country Planning Act 1971. This section of the Act allows the Secretary of State for the Environment to 'call in" certain planning applications for decision by him, if in his opinion, planning issues of more than local importance are involved. However, there were six special aspects to the inquiry when compared with other inquiries held under the same section. These were:
1. That it was held in front of a High Court Judge rather than a DOE Inspector.
2. That all parties were free to cross examine witnesses.
3. Opening and closing statements from both sides were admitted.
4. Witnesses were sworn under oath.
5. Issues of national and international significance were considered, in addition to local planning aspects.
6. Daily transcripts of the proceedings were made available.

The general procedure for a public inquiry is as follows: any person or organisation may object to a proposal within a specified objection period, which must not be less than six weeks from the date when the notice of the proposal is first published. Objectors may be represented by legal counsel if they wish, but this is not mandatory. Groups of objectors with common interests may make joint representations to the inquiry. Individuals or organisations *supporting* the proposal may also participate.

The Secretary of State for the Environment appoints an inspector, normally chosen from a corps of DOE planning inspectors, and in cases where expert knowledge in a specialist field is required, one or more assessors may be appointed. This is most common in relation to road proposals. At least six weeks notice of the commencement of the inquiry must be given.

The Inquiry at Windscale was made possible by the County Planning Authority taking the view that the proposed development formed a 'departure from a fundamental provision on the development plan'. The County Council Planning Committee though being 'minded to approve' the application were obliged to give the Secretary of State the opportunity to 'call it in'. After much public pressure, and after extending the period allowed for him to decide whether or not to call in the application he did so. This was in spite of opposition both from civil servants within his department, and ministerial colleagues on the cabinet committee which considered the matter (1).

The Inquiry commenced on 14 June 1977 and sat on each weekday until 4 November, with a short break for the preparation of final submissions. At the conclusion of the Inquiry, the Inspector and his assessors retired to write their report. This was submitted to the Secretary of State who can accept, reject or modify the recommendations in such reports as he sees fit. The Secretary of State's decision is final and there is no appeal except on a point of law.

Provision also exists for the convening of a Planning Inquiry Commission under the Town and Country Planning Act 1968 (Sections 61–63) and re-enacted in the Town and Country Planning Act 1971 (Sections 47–49). A Planning Inquiry Commission would, according to the Minister of Housing and Local Government at the time of the second reading of the Bill, deal with 'wide or novel issues of more than local significance', and would consist of between three and five members, would be empowered to employ research staff, and their procedures and terms of reference would allow them to investigate development proposals in depth, and examine alternative schemes, either at the instigation of ministers, or on their own initiative, or at the request of the developer. The Town and Country Planning Association advocated this form of inquiry for the Windscale development. However, although there has been statutory provision for a P.I.C. for over a decade, none have been convened. Giving the reasons for this Peter Shore (2) said:

'The system envisaged a two-stage procedure, the first being investigatory, and the second consisting of one or more public local inquiry. In my view the investigative proceedings are bound to lead the planning inquiry commission to conclusions, by whatever means the proceedings may be conducted. Yet at the second stage, i.e. at the local inquiry, arguments of policy and principle on which they will have already formed a view are bound to be put to them as well as the more local issues, and I do not think that people will feel that they would get a fair hearing'.

It is interesting to note that George Dobry QC for the Isle of Man Government maintained that but for the fact that BNFL's expansion proposals included buildings above 20 metres high no planning permission would have been needed as they already had blanket planning permission granted, for the previous developments on the site. It was only the height of these buildings that required BNFL to reapply for planning permission for the whole of the new development that was subsequently the subject of the inquiry.

Notes
1. See, Bugler, J. (1978) Windscale, a case study in public scrutiny *New Society* 27 July.
2. Speech, 13 September 1978.

Appendix 3. The Terms of Reference set by the Secretary of State for the Environment for the Inquiry to be conducted by Mr. Justice Parker, and the Questions that arose from them

The terms of reference were:

1. The implications for the proposed development for the safety of the public and for other aspects of the national interest.

2. The implications for the environment of the construction and operation of the proposed development in view of the measures that can be adopted under (i) the Radioactive Substances Act 1960 to control the disposal of solid gaseous liquid wastes which would result from the proposed development; and (ii) the Nuclear Installations Act 1965 to provide for the safety of operations at the reprocessing plant.

3. The effect of the proposed development on the amenities of the area.

4. The effect of the additional traffic movements both by road and rail which would result from the proposed development.

5. The implications of the proposed development for local employment.

6. The extent of the additional provision that would need to be made for housing and public services as a result of the proposed development.

From these Mr. Justice Parker outlined three basic questions which arose:

1. Should oxide fuel from United Kingdom reactors be reprocessed in this country at all, whether at Windscale or elsewhere?

2. If yes, should such reprocessing be carried on at Windscale?

3. If yes, should the reprocessing plant be about double the estimated size required to handle United Kingdom oxide fuels, and be used, as to the spare capacity for reprocessing foreign fuels?

Index

accidents
  analysis with regard to security risks of 115–22
actinides (see also alpha emitters, plutonium, americium) 100
Adams, P. 74, 77, 110, 157, 192
advanced gas cooled reactor (AGR), 81, 87, fuel cladding corrosion 174
Alesbury, A. 66, 75, 153, 194
Alexander 178, 191
Allday, C. 176, 188, 191
alpha emitters
  security and discharge control 73
  discharges 99–101, 114
  at Ravenglass 150, 161–5
  control of 165–6
americium, 99–101, 161–5, 169, 170
Armstead 58–9, 193
Armstrong-Evans 58, 193
Ashworth, G. 177, 189, 192
atmospheric discharges (see discharges)
Atherley, G. 140, 146, 148, 155, 194
Avery, D. 62, 64, 66, 69, 73, 191

Bailey, J. 179, 180–1, 191
Bainbridge 124, 192
Barratt, R. 32, 46, 102, 110, 156, 178, 187, 188, 194
Bartlett, 77, 106, 109, 124, 125, 191
Bartlett, J. 181, 193
Battersby, 191
Baynes, 177, 179, 192
Beckett, 195
BEIR Committee, 83, 84, 141, 143
Berry, G. 177, 192
Blackith, R. 48, 153, 194
Blom-Cooper, L. 79, 193
Boeck, 166, 194
Boles, 195
Bowen, 125, 159, 160, 164, 168, 169, 170, 171, 174, 185, 192
Bowers, 192
Bowie, 125, 174–5, 195
British Nuclear Fuels Ltd. (BNFL) 50, 55, 62, 64, 71, 72, 73, 78, 83, 88, 90, 91, 92, 95, 99, 105, 106, 115, 116, 118–9, 121, 125, 127, 128, 130, 139, 144, 147, 148, 149, 152, 159, 161, 166, 167, 169, 174, 176, 191
Bryant, 195

caesium 137
  unforseen increased discharge 96–9
  in fish 98, 150
cancer
  liability for public injury 77
  and the work force 101–11, 152–4
  risk assessment of 143–4, 146–50
  from fish consumption 185
Candu, 52
carbon-14, 128, 132–8, 142
Carr, 117, 195
Central Electricity Generating Board (CEGB) 56, 61–2, 161, 191
Chapman, P. 47, 49, 51, 55, 58, 62, 192
Charlesworth, 116, 118, 122, 195
Chivall, 124, 195
civil liberties, 77–9
Clelland, 94, 95, 130, 165, 167, 175, 191
coal, comparison of health effects 80, 82
Coates, I. 53, 125, 194
Cochran 64, 71, 76, 193
Collins, 157, 192
combined heat and power (CHP), 49, 56, 58
Copeland Borough Council (CBC), 179, 191
co-processing, 71
Corbet, 191
Corkhill, F. 124, 192
Cumbria County Council (CCC), 97, 106, 110, 139, 147, 162, 180, 182, 183, 191
Cumbria Naturalists' Trust (CNT) 124, 192
Dalgleish, 178, 180, 182, 191
Dalton, 125, 195
Davoll, J. 47, 194
democratic accountability, 183–6
Denson, 179, 180, 191
Department of Energy (DoEn), 45, 46–7, 61, 78, 183, 195
Department of Environment (DOE), 49, 56, 61, 195
Department of Transport (DTp), 178, 195
discharges
  atmospheric 92–3, 113, 114, 128, 166
  draft limitations 86
  history of 86–7
  limits 85

marine 87–92, 112, 165–6,
  statement of intent 126–38
  security considerations 72–3
Dixon, F. 124, 192
Dolphin, 102, 105, 106, 109, 145, 154, 155, 195
Donaghue, 72, 116, 117, 118, 119, 191
Donnison, 178, 180, 191
Doran, 191
dose
  limits 84, 141, 155
  cut-off 147
Drigg, 93, 95, 96
Dudman, A. 116, 124, 156, 183, 195
Dunster, 140, 145, 146, 147, 195
Durham County Council (DCC), 81

Edward, 194
Electrical, Electronics, and Plumbing Trades Union (EEPTU), 74, 77, 157, 192
electricity growth rate, 46
Elliot, D. 181, 193
Ellis, T. 117, 144, 150, 164, 194
energy
  alternative sources 49, 56, 57, 58, 60
  conservation 49
  demand,
    alternative forecast 48
    official forecasts 45, 46–7
  gap 47, 49
environmental impact statements, 188–9
Environmental Protection Agency, 83, 143, 147
Evans, 194

Farmer, F.R. 117, 195
Fast Breeder Reactor (FBR), 50, 70, 76, 187
Fish, 124, 195
fish consumers, 141, 148, 150, 157–8, 185
Fisheries Research Laboratory (FRL), 84 111, 168
Fitzgerald, 191
Fletcher, 125, 193
Flowers, Sir Brian 52, 167
Ford Mitre Study, 60, 80, 82
Fremlin, 22, 118, 142, 148, 180, 191
Friends of the Earth (FOE), 46, 51, 54, 60, 82, 125, 167, 192
Friends of the Earth, West Cumbria, (FOE-WC), 123, 126, 157, 192
fuel-cladding, corrosion in ponds, 97

genetic risk, 110, 141–2, 185
George, M. 193
Gilbert, 106, 109, 110
Glidewell, 77, 85, 149, 176, 177, 178, 180, 185, 189, 191
Goldsmith, E. 125, 194
Gosling, 191
Goss, 139
Greenhalgh, 125, 193
Grove-White, R. 79, 193

Hall, D. 125, 181, 185, 189, 194
Hall, Professor D. 57, 193
Halliday, 124, 192
Hanford, 105, 106
Harper, J. 27, 170, 192
Hatton, G. 124, 195
Haworth, C. 123, 158, 192
Health and Safety Executive (HSE), 156
Henderson, 124, 195
Hermiston, 87, 97, 111, 142, 143, 156, 161, 162, 191
Herzig, 66, 78, 195
Hetherington, 174
high active waste (HAW) tanks, 128, 130
  accidental release 116–22
  sabotage 121
  security 72
Highman, M. 124, 156, 181, 195
Hillier-Fry, 124, 195
Holden, I. 124, 195
Hookway, R. 195
Hopper, 176, 178–9, 180, 191

Ichikawa, 110, 142, 146, 171, 193
Industrial action, restriction, 73, 74
inquiry
  costs of 33, 187, 189
  procedure 187–8
iodine-129, 128, 132–8
International Commission on Radiological protection (ICRP), 83, 84, 105, 144, 127, 128, 130, 132, 140, 141, 143, 144–5, 147, 148, 149, 150, 152, 158, 171, 172, 174, 184
International Atomic Energy Authority (IAEA), 69, 70, 83, 122, 123, 148, 184
Isle of Man Local Government Board (IOM), 81, 124, 150, 166, 168, 170, 171, 173, 192

Jenkins, 58, 193
Johnston, 65, 192
Jones, 124, 195
Jones, T.P. 45, 57, 60, 61, 183, 195
Justice, 76, 78, 192

Kidwell, R. 46, 50–1, 52, 54, 55, 60, 61–2, 64, 66, 75, 77, 85, 181, 189, 192
Kneale, G. 105, 106, 109, 110, 139, 194
krypton-85, 128, 132-8, 147, 161, 166

Lake District Special Planning Board (LDSPB), 177, 179, 192
Lancashire and Western Sea Fisheries Joint Committee (LWSFJC), 82, 172, 185, 192
Layfield, Sir Frank 176, 194
Laxen, D. 150, 155, 162, 193
Leach, G. 36, 46–7, 49, 56, 192
Lewis, R. 74, 77, 193
Liddle, 174, 195
limiting environmental capacities, 150, 152, 160, 164
Lindop, P. 140
Little, K. 23, 125, 193
Local employment potential, 180–1
Macdonald, 125, 193
Macfadyen, D. 194
McLean, 140, 145, 146, 154, 195
McLeod, 124, 192
Magnox, problems with fuel corrosion, 97
Mancuso, 105, 139
Maxwell, 149
Medical Research Council (MRC), 83, 114 144, 150, 169, 170
Miller, 124, 195
Milne, 75, 116, 191
Ministry of Agriculture, Fisheries and Food (MAFF), 81, 112, 126, 151, 158, 160, 161, 162, 166, 169, 174, 184, 195
Mitchell, 160, 169, 170, 195
Morgan, 143
Morley, 170, 195
monitoring programmes, 81, 85, 151, 159, 161–5
mixed oxide fuel (MOX), 76
Mummery, P. 141, 149, 153, 157, 191
Musgrove, P. 57, 60, 193
Murray, 191

National Council for Civil Liberties (NCCL), 79, 193
National Peace Council (NPC), 193
National Radiological Protection Board (NRPB), 83, 102, 105, 109, 123, 139, 140, 143, 144, 145, 147, 152, 153, 154, 155, 160, 161–5, 169, 170, 171, 172, 184, 186, 195
Natural Resources Defense Council (NRDC) 76, 193

Naylor, 192
Network for Nuclear Concern (NNC), 96, 123, 146, 150, 160, 162, 165, 168, 170–4, 187, 193
Niven, 172, 195
Non Proliferation Treaty (NPT), 64, 65, 69
niobium-95, 99
Norman, 124, 192
Nuclear Installations Act 1963, 77, 84
Nuclear Installations Inspectorate (NII), 81, 172, 195

Oakes, S. 64, 66, 193
Odell, P. 47–9, 194
O'Riordan, 195

Page, 48, 51, 57, 193
Parker, Mr. Justice
  comments on:
    accident risk 121, 147–8
    B 204/205 plant, use of, 54
    civil liberties 74, 76, 79
    employment 181, 182
    energy, alternative sources 60
    energy conservation 59
    energy demand 46, 48, 49
    Environmental Impact Analysis 189
    financial aspects of THORP 62
    financial disparity of participants 184
    infrastructure 197
    inquiry procedure 187
    plutonium, safeguarding of, 70
    public accountability, and acceptance of control 172, 173, 174, 184, 186
    public hostility to nuclear power 125–6
    radioactive waste, control of, 82
    radioactive waste, storage of, 52
    radioactive waste, vitrification of, 175
    research and organisation 170
    resources diplomacy 55
    site selection 168, 176–7
    wet fuel storage 167–8
    visual impact 177
  requests:
    consequence studies for loss of cooling 119
    sampling of Ravenglass air 162, 164–5
    tests on local fish consumers 150
    whole body monitoring 157–8
  pathways, 111–5, 151–2, 168–71
Patterson, W. 58, 192
Paulin, D. 124, 193
Pedlar, K. 75, 194

plutonium (see alpha emitters)
  activity in air at Ravenglass, 114, 161–65
  activity in air 114, 161–65
  behaviour in lung 144
  control of, in discharges, 165
  reactor grade and nuclear weapons 75
  transfer factors 139, 169, 174
plutonium economy, 77, 78, 79
Pochin, Sir Edward 102, 140, 141, 142, 145, 146, 153, 154
Political Ecology Research Group (PERG) 72, 73, 78, 96, 118–9, 122, 125, 146, 165, 168, 170, 173–4, 184, 185, 187, 193
Postlethwaite, 123, 191
Potts, 142, 164, 185, 192
Price, 51, 194
public hostility, 122–6
public participation, 33–5, 186, 187, 189

Quayle, 192

Radford, 125, 140, 142, 144, 145, 146, 147, 155, 162, 164, 170, 171, 174, 193
radiation, workers' exposure to, 101–11, 152–7
radioactive waste disposal, 82, 93, 165–8,
  high active liquids 94
  high active solids 94
  low active 95
  medium active 94–5
radon gas, health effects 80
Randell, 58, 193
Rasmussen Report, 120
Ravenglass, 150, 161–5, 169
Rich, M. 106, 178, 191
resource diplomacy, 55
Richardson, 124, 195
Rippon, 125, 193
rish assessment
  of disease 146–50
  of accidents 116–26
Robertson, 74, 116, 195
Robinson, 192
Rosenthal, 195
Rotblat, J. 52, 65, 67, 70, 105, 110, 194
Royal Commission on Environmental Pollution
  Government response to: 83
  Sixth report: Nuclear Power and the Environment, 66, 75, 77, 79, 81, 83, 85, 94, 111, 115, 140, 148, 152, 153, 156, 169, 170, 171, 188
ruthenium-106, 99

sabotage, 75
safety, 115–22
  and strike action 73, 116
Salter, 57, 193
Scargill, A. 57, 59, 193
Schofield, 102, 139, 153, 154, 156, 191
Scott, A.I. 176, 181, 191
Scott, 195
seaweed, 99, 160
security, 72–6
  and the control of technology 72
  and HAW tanks 72
  discharge control 72–3
  and the rights of the workforce 73–4
  and terrorism/sabotage 75
Sedley, S. 193
Sefton, 178, 192
Shaw, 195
Shorthouse, 174, 194
Shortis, 74, 105, 127, 191
Sieghart, P. 77, 78–9, 192
Silkwood Case, 75
Silsoe, Lord 20, 45, 49, 52, 55, 56, 57, 58, 60, 64, 99, 103, 106, 110, 150, 156, 180, 191

Sly, 124, 195
Small, 195
Smith, 72, 75, 154, 191
Social impact, 123, 125
Socialist Environment and Resources Association (SERA), 73, 193
Society for Environmental Improvement (SEI) 57, 125, 193
South of Scotland Electricity Board (SSEB) 62, 194
Spearing, J. 80, 85, 105, 124, 141, 142, 146, 194
spent fuel storage, 82
  corrosion problems in water 167
  dry storage 167
  leach rates 167
Spencer, Sir Kelvin, 125, 194
Stewart, A. 105, 106, 109, 110, 139, 143, 154, 155, 194
Stoel, T. 168, 188, 194
Stredder, R. 195
subversion, 78
surveillance, 78
Sweet, C. 62, 194
Swift, 179, 191
synergism, 160

Taylor, P. 80, 82, 117–8, 121, 122–3, 125, 140, 145, 146, 148–9, 159, 160, 161, 165–6, 169–70, 174, 183–4, 193
terrorism, 75, 76, 121
Thirlwall, G. 168, 177, 188, 194
Thompson, G. 72, 116, 120–1, 185–6, 193
Thompson, J. 88, 92, 96, 97, 99, 150, 169, 173–4, 193
Thorold, O. 36, 55, 192
TIRION Computer programme, 117, 118, 119, 120, 187
Tolstoy, I. 36, 193
Tosswill, 195
Town and Country Planning Association (TCPA), 96, 105, 110, 140, 168, 187, 194
Trades Union Congress (TUC), 73, 74, 77, 125
transport
  of nuclear fuel 75
  of plutonium 75
tritium, 142, 165–6
Tremlett, 124, 195
de Turville, 60, 193
Tweedy, 57, 194
Tyme, J. 58, 193

United Kingdom Atomic Energy Authority (UKAEA), 52, 78–9, 87, 175, 195
UNSCEAR, 141, 160
uranium, 46, 50–1, 52, 55
Urquhart, J. 96, 97, 100, 142, 194

vitrification, 82, 121, 167, 174–5

Wadsworth, U. 124, 195
Wakstein, C. 115, 126, 194
Warner, Sir Frederick, 19
Warner, 73, 92, 94, 97, 99, 100, 126, 127, 128, 166, 167, 172, 174, 191
Webb, 147, 186, 195
West Germany
  dose limits 141
  Nordrein Westfalen moratorium 185
Widdicombe, D. 35, 36, 50, 56, 58, 68, 75, 172, 174, 194
Windscale Appeal (WA) 75, 76, 173, 187, 194
Windscale Assessment and Review Project, 33
Windscale Inquiry Equal Rights Committee (WIERC), 73, 194
Williamson, 191
Wilson, 116, 191

Wilson, Lord 57, 193
Wilson, Professor, E. 57, 193
Wohlstetter, 64, 66–8, 70, 71, 192
Wright, 45, 56, 61, 62, 191
Wynne, B. 105, 110, 124, 145, 146, 150, 155, 170, 193

zirconium-95, 99